DEVELOPMENTS IN FOOD SCIENCE 14

MOLECULAR GENETICS
An Outline for Food Chemists and Biotechnologists

DEVELOPMENTS IN FOOD SCIENCE

Volume 1 J. G. Heathcote and J. R. Hibbert
Aflatoxins: Chemical and Biological Aspects

Volume 2 H. Chiba, M. Fujimaki, K. Iwai, M. Mitsuda and Y. Morita (Editors)
Proceedings of the Fifth International Congress of Food Science and Technology

Volume 3 I. D. Morton and A. J. MacLeod (Editors)
Food Flavours. Part A. Introduction; Part B. The Flavour of Beverages

Volume 4 Y. Ueno (Editor)
Trichothecenes: Chemical, Biological and Toxicological Aspects

Volume 5 J. Holas and J. Kratochvíl (Editors)
Progress in Cereal Chemistry and Technology. Proceedings of the VIIth World Cereal and Bread Congress, Prague, Czechoslovakia, 28 June—2 July 1982

Volume 6 I. Kiss (Editor)
Testing Methods in Food Microbiology

Volume 7 H. Kurata and Y. Ueno (Editors)
Toxigenic Fungi — Their Toxins and Health Hazard. Proceedings of the Mycotoxin Symposia held in the Third International Mycological Congress, Tokyo, Japan, 30 August—3 September 1983

Volume 8 V. Betina (Editor)
Mycotoxins. Production, Isolation, Separation and Purification

Volume 9 J. Holló (Editor)
Food Industries and the Environment. Proceedings of the International Symposium, Budapest, Hungary, 9—11 September 1982

Volume 10 J. Adda (Editor)
Progress in Flavour Research 1984. Proceedings of the 4th Weurman Flavour Research Symposium, Dourdan, France, 9—11 May 1984

Volume 11 J. Holló (Editor)
Fat Science 1983. Proceedings of the International Society for Fat Research Congress, Budapest, Hungary, 4—7 October 1983

Volume 12 G. Charalambous (Editor)
The Shelf Life of Foods and Beverages. Proceedings of the 4th International Flavour Conference, Rhodes, Greece, 23—26 July 1985

Volume 13 M. Fujimaki, M. Namiki and H. Kato (Editors)
Amino-Carbonyl Reactions in Food and Biological Systems. Proceedings of the 3rd International Symposium on the Maillard Reaction, Susuno, Shizuoka, Japan, 1—5 July 1985

Volume 14 J. Škoda and H. Škodová
Molecular Genetics. An Outline for Food Chemists and Biotechnologists

DEVELOPMENTS IN FOOD SCIENCE 14

MOLECULAR GENETICS

An Outline for Food Chemists and Biotechnologists

J. ŠKODA

Institute of Organic Chemistry and Biochemistry of the Czechoslovak Academy of Sciences, Prague, Czechoslovakia

H. ŠKODOVÁ

Research Institute for Biofactors and Veterinary Drugs, Pohoří-Chotouň, Czechoslovakia

ELSEVIER

Amsterdam—Oxford—New York—Tokyo 1987

Published in co-edition with SNTL – Publishers of Technical Literature, Prague

Distribution of this book is being handled by the following publishers:

for the U.S.A. and Canada
ELSEVIER SCIENCE PUBLISHING COMPANY, INC.
52 Vanderbilt Avenue, New York, N. Y. 10017

for Albania, Bulgaria, Chinese People's Republic, Cuba, Czechoslovakia, German Democratic Republic, Hungary, Korean People's Democratic Republic, Mongolia, Poland, Romania, the U.S.S.R., Vietnam and Yugoslavia
SNTL – Publishers of Technical Literature
Spálená 51, 113 02 Praha 1, Czechoslovakia

for all remaining areas
ELSEVIER SCIENCE PUBLISHERS
Sara Burgerhartstraat 25
P. O. Box 211
1000 AE Amsterdam, The Netherlands

Library of Congress Cataloging in Publication Data

Škoda, J. (Jan)
Molecular genetics.

(Developments in food science; 14)
Translation of: Molekulární genetika.
Bibliography: p.
Includes index.
1. Molecular genetics. 2. Biotechnology. 3. Food – Analysis. 4. Food – Composition. I. Škodová, H. (Helena) II. Title. III. Series.
QH506.S55 1987 574.87'328 86-13378
ISBN 0-444-99500-5 (Vol. 14)
ISBN 0-444-41688-9 (Series)

With 81 illustrations and 10 tables

Printed in Czechoslovakia

Contents

Preface

Dear reader,

We would like to invite you to take part in an adventurous trip to the fantastic microworld of informational molecules which contain the code of life, including your own. We want to lead you nearer to the complicated and incredibly accurate work of a living cell in which these informational molecules are formed and where they control and steer its biological functions through a marvellously ingenious mechanism. We invite you to learn to understand the molecular basis of heredity, the discovery of which rates among the most significant achievements of modern science. We shall guide you along a path beginning with the biosynthesis of informational molecules and the laws governing their structure, all the way to the reproduction and functions of these magnificent natural polymers; in their company you will become acquainted with the recent brave attempts of molecular geneticists to interfere with the heredity of living organisms – genetic engineering.

It is essential to realize at the very beginning how minuscule these informational macromolecules are. If an average man's height were equal to the size of Great Britain the informational macromolecule (a double-stranded deoxyribonucleic acid), which is highly economically located in every somatic cell of the super-giant, would appear as a filament about 1 mm in diameter and several hundred kilometres long! The existence of man and all his faculties – as well as those of all other forms of life – are thus based on an incredibly tenuous matrix.

Even the concept of time is different in this micro-cosmos. The singularly complex molecule of a protein formed by several hundred building blocks of twenty different kinds and associated in a precise sequence can be formed in a living cell within a few seconds. Once it is grasped by what complicated and multiply regulated mechanisms a protein molecule is formed, it is almost beyond human comprehension to appreciate the speed at which these intricate and involved processes take place. You will soon appreciate the fact that the apparent perfection of life on the macro-scale is made possible by a no less perfect and admirable micro-world of macromolecules. You will recognize and accept that life is not based on a simple principle, some "living primeval substance", but that the complexity of the biological macro-cosmos only reflects complicated motion in the micro-cosmos of informational molecules. On the other hand, it is a highly satisfying thought to realize that the fundamental principles of all life on Earth are the same at the level of these informational molecules, being based on the existence of very similar, even identical polymers. This conclusion is unequivocal proof of the origin and evolution of life on Earth from the same principle.

The road from the beginnings of classical genetics to molecular genetics was very long and tortuous, fraught with obstacles. Classical genetics needed a whole century to reach definitive conclusions. Mendel is recognized as its founder, the Mendel who in 1865 formulated his classical laws on the basis of experiments with genetic crosses of the pea (*Pisum sativum*). Mendel's genetics localized the hereditary traits into the so-called alleles which occur in pairs. These paired alleles segregate during the formation of gametes (male and female sexual cells) and are recombined during the fertilization process. The model derived from Mendel's studies has a purely statistical character. The segregating units were called genes in 1911.

Morgan's experiments on the fruit-fly (*Drosophila melanogaster*) extended Mendel's laws of heredity by important facts. Genes were found to be organized in certain groups that show mutual linkage during genetic transfers. Within each linkage group genes are ordered in a sequence that can be expressed by a linear genetic map. In such a map short distances between any two genes will result in a small recombination probability whereas longer distances will lead to more frequent recombination during crossing. Morgan's experiments led to the important conclusion that the number of linkage groups exactly matches the number of chromosome pairs of somatic (body) cells.

Another important factor that enriched and extended classical genetics was the discovery of mutation (de Vries); physical or chemical factors were shown to induce changes in hereditary properties.

The onset of the so-called biochemical genetics was marked by experiments with microorganisms and later with phages. When mutagenic factors are permitted to act on microbial cells mutants may arise that do not contain an enzyme, which is essential for the formation of a substance required for growth. Such mutants will grow only if the requisite substance is added to their growth medium. It is through the use of such deficient mutants that a great many problems in biochemistry and "pure" genetics have been solved.

Results of studies of classical genetics led to the conclusion that living matter can be reproduced only if suitable instructions, located in the chromosomes, are present. These instructions, subject to well-defined rules of genetic transfer, are instrumental in the formation of cell proteins (enzymes) and can be influenced by mutagens. However, the chemical nature and structure of this genetic material remained unresolved.

It was in 1928 that Griffith discovered bacterial transformation. He was able to demonstrate that virulence (infectivity) of pneumococci inactivated by heat can be transferred to a nonvirulent (noninfective) strain. Three years later such a bacterial transformation was accomplished *in vitro* by Alloway. After several years of analysing the transformation process Avery and co-workers could declare that the only cell fraction with transformation activity is the fraction of deoxyribonucleic acids. Identification of deoxyribonucleic acids as genetic material and the elucidation of their double-helical structure by Watson and Crick in 1953 laid the foundation of molecular genetics.

In studies of the molecular genetics of microorganisms, bacteria and fungi have been used. However, bacterial mutants are increasingly being employed in production microbiology. Genetic engineering, permitting transfers of genes to be made between different species, opens fundamentally new perspectives for fermentation processes.

Progress in molecular genetics and molecular biology of the cell provides a new impetus for studies of cell differentiation and of the basis of tumour growth. An excellent and uniquely useful model for such studies was found in viruses and the viral transformation of cells. Even immunology is positively stimulated by molecular genetics.

In preparing this volume, we took upon ourselves a complicated task: to select from the enormous breadth of contemporary molecular genetics of microorganisms only the very fundamental information that is germane to the productive development of food chemistry and biology. We assumed that the reader would have a solid knowledge of basic chemistry but only of the principles of biochemistry and biology – hence we attempted to compile something like basic genetics for food chemists and biotechnologists. We were aided in this work by many years of experience in teaching at the Institute of Chemical Technology in Prague. As we were convinced that graphical clarity is of primary importance for understanding the more complicated relationships and for retaining them in the memory, the text is accompanied by a number of schematic representations and illustrations.

We are indebted to the reviewers of the manuscript, Professor Oldřich Nečas and Dr. Václav Pačes, for many valuable comments, to Mr. František Zemanec for collaboration in the illustrations and last but not least to Dr. Arnošt Kotyk for his professional translation of the book into English.

J. Škoda and H. Škodová

Heredity

Heredity in the broad sense is the biological continuity between parents and their progeny. Genetics studies the basis, transmission and variability of hereditary properties.

The basic functional unit of genetic analysis is the gene. It is a segment of deoxyribonucleic acid with a specific number and sequence of deoxyribonucleotides and with a specific function.

From the genetic point of view, every cell is an independent unit with its own genotype that represents the complete set of its genetically coded potentialities (genes). The expression of these potentialities is observed as biochemical or morphological characteristics – traits. The genotypes of all individual cells of a multicellular organism are usually identical.

The overall properties of the organism arising through the interaction of the genotype with the environment of the cell or the organism are called the phenotype. Phenotype is thus the expressed part of the genotype and it depends not only on external conditions but is influenced by numerous internal factors controlled by the complex organization of informational macromolecules.

The phenotypic expression of the organism is made possible by a genetic transcription of the information stored in the chemical structure of deoxyribonucleic acid (i.e., in the genotype) into the structure of ribonucleic acid, and by a genetic translation of this transcript into the structure of enzymes and other proteins. The genetic code, according to which genetic information is translated, is basically the same for all organisms.

Nucleic Acids

Introduction

Genetic information in both unicellular and multicellular organisms is stored in deoxyribonucleic acids (DNA) or, in some viruses and phages, in ribonucleic acids (RNA). This conclusion is supported by the following findings:

a) Microbial DNA freed of other macromolecules can transfer into a cell either a single gene, a group of genes or all the genes of a chromosome.

b) Diploid somatic cells of higher organisms always contain the same amount of DNA of the same chemical composition. The amount of DNA in a cell increases during interphase between mitotic divisions (p. 112) but returns to the original constant value in daughter cells. Sexual (i.e., haploid) cells contain DNA of the same composition as the diploid ones, but in half the amount.

c) Physico-chemical and chemical mutagens alter the structure of DNA.

d) DNA represents the matrix on which RNA is copied or transcribed, the resulting RNA determining the sequence of amino acids during protein synthesis (translation of genetic information).

e) The information capacity of DNA per cell is sufficiently large.

It might thus be taken as established that the material basis of heredity is DNA in which genetic information is stored by the so-called genetic code. The information capacity of DNA is fantastic – in a human germ cell all the information needed for the function of the human organism is already present. If it were possible to obtain DNA from the germ cell of every human on Earth, the total biological programme of mankind would be available in the size of a small pill.

The historical development of information about nucleic acids may be divided into three periods. The first of these begins with the discovery of nucleic acids, made in 1869 by the Swiss physician Miescher as he was examining the chemical composition of cells. Miescher worked with leukocytes isolated from pus because he thought that these were the simplest animal cells. One of the tasks he wanted to tackle was the isolation of cell nuclei (the cell nucleus had been discovered in 1851). On applying hydrochloric acid to leukocytes, Miescher obtained nuclear material which he then freed of protein with the aid of pepsin. The material was extracted with alkali and re-precipitated with acid. It was thus apparently an acid for which Miescher coined the name nuclein. Later Miescher isolated nuclein from salmon sperm, the cells of which contain more than 90% of nuclear material. He showed nuclein to be combined in the nuclear material with a basic protein which he called protamine. It was demonstrated only later that nuclear DNA may be combined with basic proteins of two types, the protamines or the histones, and thus form the so-called nucleoprotein.

Meischer's work was taken up by a number of scientists. The term "nucleic acid" appears first in the work of Altmann in 1899. Picard isolated purine bases, then Kossel pyrimidine bases from nucleic acids.

It was during the early stages of nucleic acid research that it became apparent that there are two types of "nuclein". One was isolated from yeast and its acid hydrolysis yielded four nitrogen-containing bases (adenine, guanine, cytosine and uracil), phosphoric acid and a pentose, identified as D-ribose. The second type of nucleic acid was prepared from the thymus and, after intense acid hydrolysis, it gave rise to four nitrogen-containing bases (adenine, guanine, cytosine and thymine), phosphoric acid and 2-deoxy-D-ribose. The first type was called ribonucleic acid (RNA) and the second deoxyribonucleic acid (DNA).

The analyses carried out subsequently gave rise to hypotheses about the structure of nucleic acids. Up to about 1940 the so-called tetranucleotide hypothesis had generally been accepted, based on an inaccurately determined molecular weight of RNA ($1.3 \, . \, 10^3$). According to this hypothesis both RNA and DNA were composed of tetranucleotide subunits formed by a combination of the four nucleotides containing adenine, guanine, cytosine and uracil (or thymine). The tetranucleotide hypothesis was subsequently shown to be erroneous and more refined extraction techniques made it possible to isolate various preparations of RNA and DNA of molecular weights ranging from $2 \, . \, 10^6$ to $8 \, . \, 10^6$. As was shown recently, eukaryotic chromosomes consist of single DNA chains of molar mass as large as $5 \, . \, 10^{10}$.

The second period in the research of nucleic acids was characterized by an analytical approach to the problem. The development of paper chromatography, ion-exchange chromatography and electrophoresis of different types made it possible to separate and identify the degradation products of nucleic acids. This period then culminated in an investigation of the conformational structure of polynucleotide chains. It was in 1953 that Watson and Crick published their history-making concept of DNA structure. The fine structure of DNA was further elucidated by a number of workers, including Chargaff, Franklin and Wilkins. At that time, nucleic acids were already assumed to be carriers of hereditary properties. The experimental verification of the Watson-Crick model of DNA structure concludes the second period of nucleic acid research.

The third period includes the enzymic synthesis of nucleic acids and their components. It became possible to synthesize nucleotides, oligonucleotides and polynucleotides by progressive methods of organic chemistry. The molecular basis of transmission of genetic information from DNA to RNA to proteins was unravelled and the largely universal code through which the information is maintained was deciphered.

During this, still unterminated, period the study of nucleic acids is undergoing explosive development, its consequences continuously enriching a number of scientific disciplines, especially molecular genetics.

Nucleic acids hold a key position in tackling various genetic problems, so consideration will now be given to the structure and biosynthesis of these informational macromolecules.

Structure of DNA

DNA forms long unbranched polymers, their backbone being composed of regularly alternating residues of 2-deoxyribose and phosphoric acid, linked by ester bond. Each deoxyribose residue then binds a nitrogenous base [1]. The structure of the chain is shown by scheme [2]. The same segment of the chain of DNA can be expressed by formula [3].

[1]

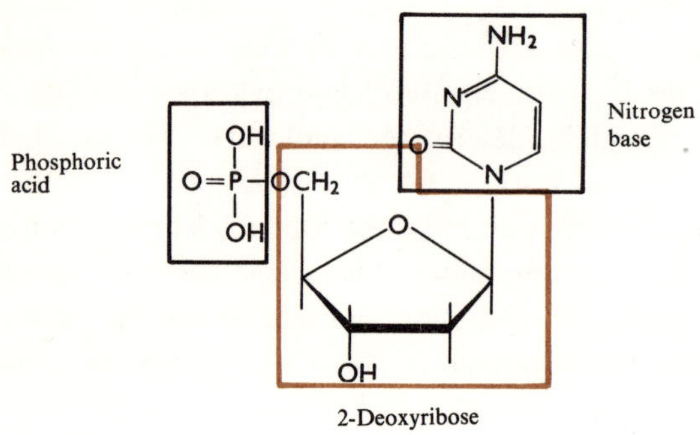

[2]

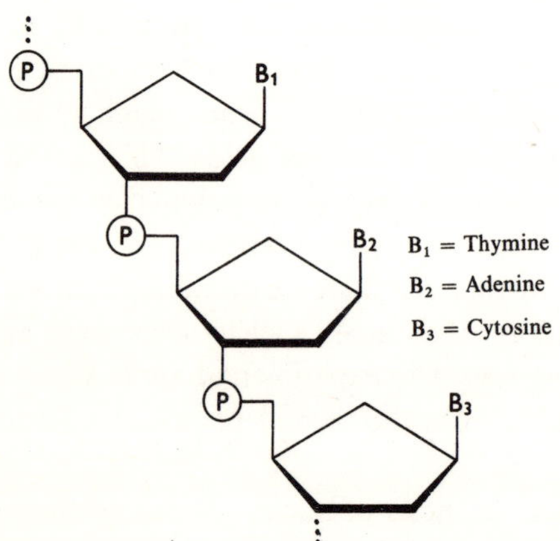

Monomeric components of DNA are deoxyribonucleotides, containing a nitrogenous base, 2-deoxy-D-ribose, and phosphoric acid [1].

Bases isolated from DNA are of the purine or pyrimidine type. The numbering of atoms in purine and pyrimidine rings is shown in the formulas of adenine and uracil [4].

[3]

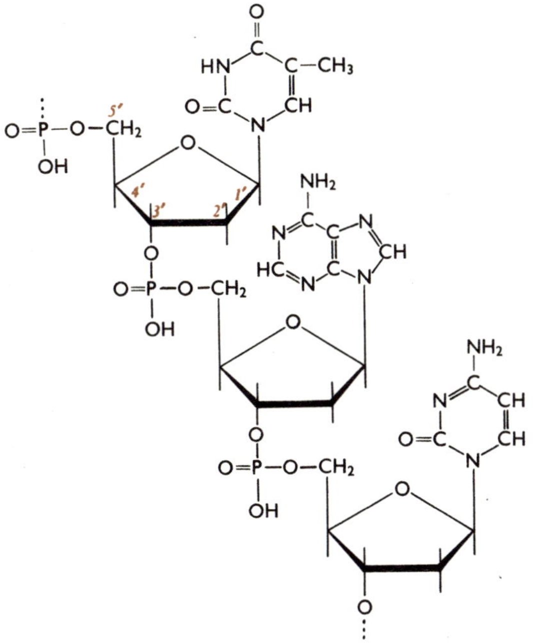

[4]

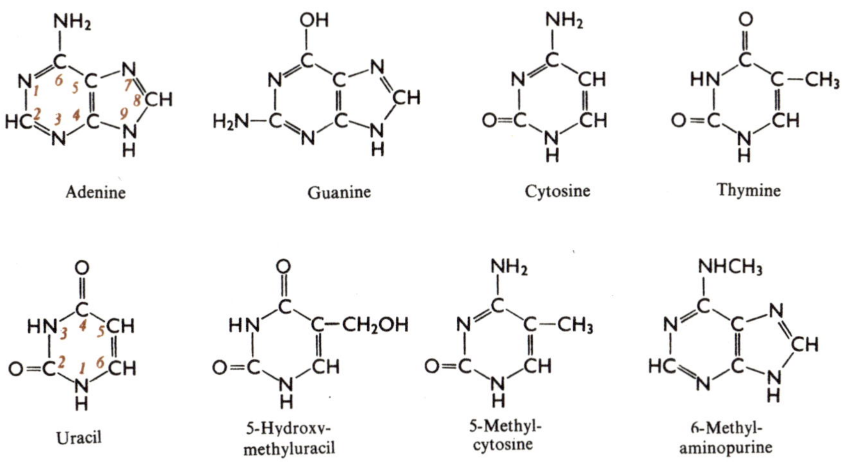

In addition to the purines adenine and guanine, and the pyrimidines cytosine and thymine, DNA may contain some less common bases – for formulae see [4]. DNA of some bacteriophages contains uracil and 5-hydroxymethyluracil, DNA of the "T-even" bacteriophages contains 5-hydroxymethylcytosine instead of cytosine. Some animal and bacterial DNAs contain 5-methylcytosine and fairly large amounts of this base are present in the DNA of higher plants. A minor component of some bacterial and phage DNA is 6-methylaminopurine.

2-Deoxyribose, the sole sugar component of DNA, forms deoxyribonucleosides bonded to a nitrogenous base. An ester bond at carbon 5′ of the deoxyribose residue with phosphoric acid then gives rise to deoxyribonucleotides – see formula [1].

As follows from the formula of the DNA segment on p. 17, the nitrogenous bases are always bound to deoxyribose carbon 1. The phosphoric acid residues then join deoxyribose carbon 5′ of one deoxyribonucleoside with a carbon 3′ of the neighbouring deoxyribonucleoside. This is the basis of what is called the 3′→5′ internucleotide (phosphodiester) bond.

The usual arrangement of the DNA molecule is that of two strands intertwined, forming the so-called double helix about 2.0 nm thick. Every turn of the double helix is 3.4 nm long and is formed by 10 nucleotides. The nitrogenous bases point towards the centre of the double helix and are joined by hydrogen bonds (Fig. 1). It is a general rule here that thymine pairs with adenine, and cytosine with guanine.

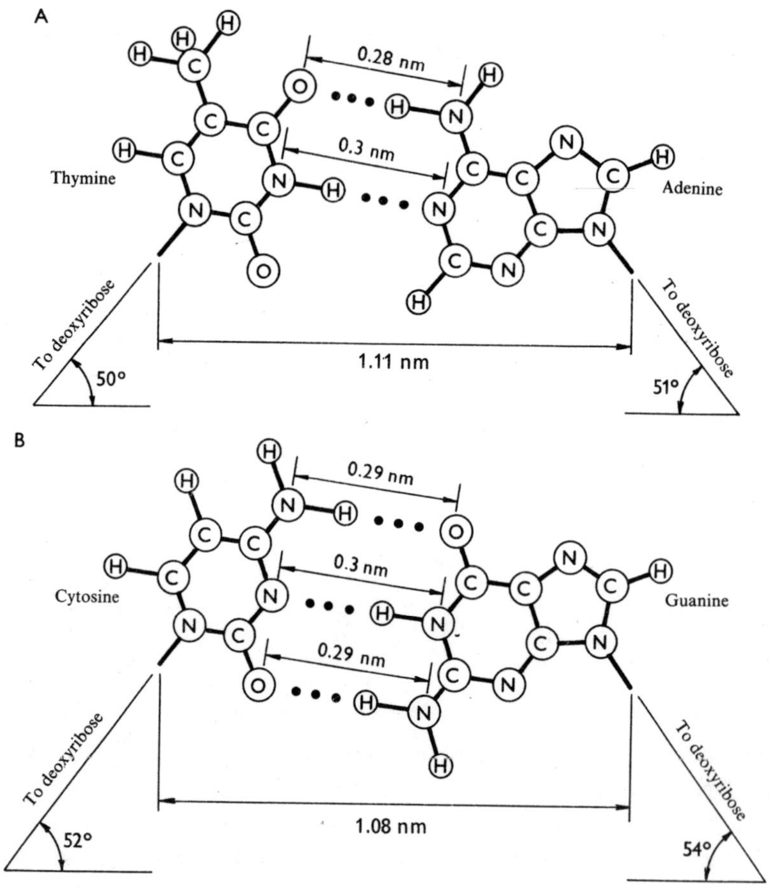

Fig. 1. Hydrogen bonds of a DNA double helix.
A, Hydrogen bonds between thymine and adenine;
B, hydrogen bonds between cytosine and guanine

However, hydrogen bonds are not the only factor that stabilizes the double-helical structure. The position of the neighbouring residues in a DNA chain permits the occurrence of van der Waals interactions, which reinforce the longitudinal stacking of DNA.

An important method for determining the structure of DNA was X-ray diffraction, which provided information on the spatial distribution of same atoms in DNA preparations. The interpretation of X-ray and chemical analysis data resulted in the proposal of the right-handed B-double-helical organization of DNA known as the Watson-Crick model.

The model of right-handed B-form DNA double helix was uncontested until 1972 when the uniformity of the Watson-Crick concept of DNA structure suffered a crack. Synthetic double-stranded polydeoxynucleotides at high concentrations of salts or alcohol were found to display unusual circular dichroism spectra. The explanation of the deviations led to the discovery of a left-handed Z configuration (Z-DNA, for zig-zag). It was shown later that the Z configuration of DNA can exist

Schematic representation of base pairing in two complementary DNA strands

Schematic representation of the double-helical form of the same segment of DNA

Molecular model of the same DNA segment

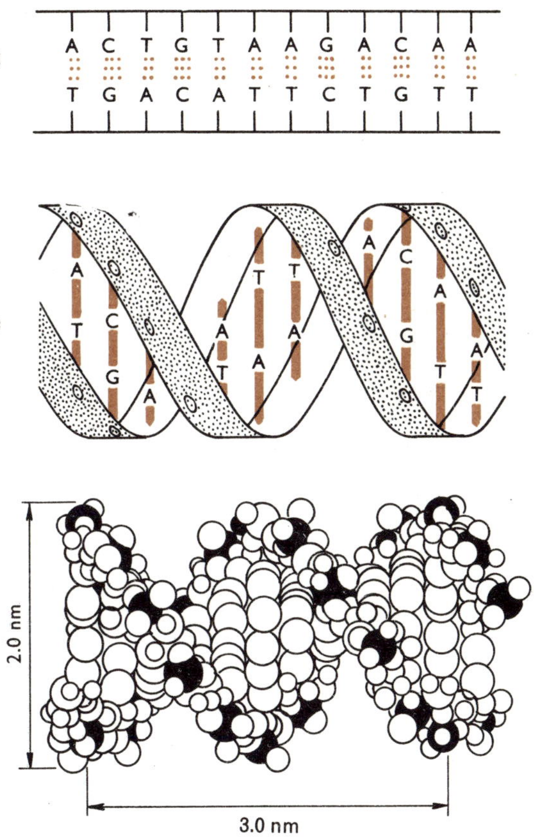

Fig. 2. Different ways of illustrating a segment of DNA double helix

not only under the above extreme conditions but that, for instance, methylation of cytosine residues in position 5 of the pyrimidine ring and other factors, such as supercoiling of DNA, facilitate the transition of the B-form to the Z-form. Hypotheses exist that the left-handed Z-form can represent that site of the DNA molecule where regulatory proteins are bound, as well as the site of replication and recombination. The Z-form makes it possible for some bases in DNA to become accessible to interactions, e.g. with proteins.

For the double helix to be regular it is essential that a purine residue be paired with a pyrimidine residue. Considering the fact that a thymine is always paired with an adenine, and a cytosine with a guanine, complementarity of the juxtaposed bases in every DNA segment becomes obvious (Fig. 2).

While the ratio of adenine to thymine and the ratio of guanine to cytosine must be identically equal to one, the proportion of the A-T and G-C pairs will differ from one DNA preparation to another. This ratio is usually expressed by (A+T)/(G+C) (Table I). DNA with a higher G+C content are more heat-stable; during thermal denaturation of DNA solutions of this type the hydrogen bonds between the two strands are severed only at higher temperatures. The reason for this lies in the fact that G−C pairs are joined by one more hydrogen bond than are the A−T pairs.

Table I. (A + T)/(G + C) ratios in DNA from various sources

Source of DNA	(A + T)/(G + C)
Phage λ	1.06
Phage T2	1.84
Escherichia coli	0.97
Saccharomyces cerevisiae	1.50
Streptomyces griseus	0.35
Aspergillus niger	1.00
Tetrahymena pyriformis	2.33
Wheat	1.22
Trout	1.34
Chicken	1.36
Horse	1.33
Man	1.40

Some viruses and bacteria contain single-stranded DNA; obviously, here the A/T and G/C ratios need not be equal to one. The bacterial chromosome is often circular.

Structure of RNA

Like in DNA, the basic primary structure of RNA is a linear polynucleotide, its backbone being composed of regularly alternating residues of phosphoric acid and ribose (as D-ribofuranose). The sugar residues then carry purine and pyrimidine bases. No circular form of RNA is known.

The monomeric components of RNA are ribonucleotides joined, in analogy with DNA, by 3′ → 5′ phosphodiester bonds. The principal bases of RNA are adenine, guanine, cytosine and uracil. Transfer RNA (tRNA) may contain an appreciable number of modified minor bases, such as hypoxanthine, 5,6-dihydrouracil and methylated bases. Transfer RNA may also contain thymine but of different biogenesis from that in DNA. An interesting component of tRNA molecules is 5-ribosyluracil (pseudouridine, ΨUrd), in which the usual N-glycosidic bond is replaced with a C–C bond [5].

[5]

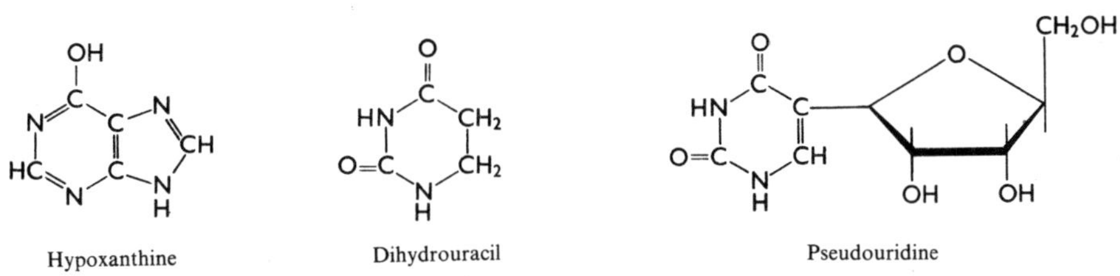

RNA may occur as an amorphous substance in which the base residues are not oriented specifically. This is the case with polyuridylic acid, poly(U), an enzymically prepared model of RNA. In some polyribonucleotides the base residues have a tendency to interact and are then located perpendicularly to the polymer backbone. As purine and pyrimidine bases are all planar molecules, a single-stranded helix can arise. Such conformations have actually been observed in enzymically prepared polyribonucleotides, such as poly(A). Under some conditions, however, RNA as well as synthetic polynucleotides can generate a double helix based on the above described pairing of bases in the juxtaposed strands. The last type of secondary structure of RNA is the hairpin conformation, which arises from a single strand of RNA coiled into a double helix (Fig. 3). Transition from a single-stranded RNA to a double-helical RNA forming a hairpin depends on the base sequence in the polyribonucleotide, temperature, pH and the ionic strength of the solution.

According to its function, RNA may be classified into four groups:
1. messenger RNA (mRNA);
2. ribosomal RNA (rRNA);
3. transfer RNA (tRNA);
4. viral RNA.

1. mRNA. The length of mRNA chains is determined by the size of the protein which it codes. If an mRNA molecule contains information for more than one protein it is called polycistronic mRNA. The molecular weight of mRNA can thus vary over a wide range. Hence, the molecular weight of mRNA formed on the His operon of *Escherichia coli,* which contains 15 cistrons, will reach a value of $4 \,.\, 10^6$, which corresponds to a length of 3 μm and a total number of 13 000 nucleotides. Eukaryotic mRNA contains a poly (A) sequence at the 3′-terminus. Many cellular and viral mRNAs are terminated at the 5′ end by a sort of cap, formed by positively charged 7-methylguanosine 5′-triphosphate; furthermore, one or two adjacent ribose residues are methylated in position 2′. This termination of the polynucleotide is thought to play two roles: (a) facilitation of mRNA binding to the ribosome (b) protection of mRNA against 5′-exonucleases.

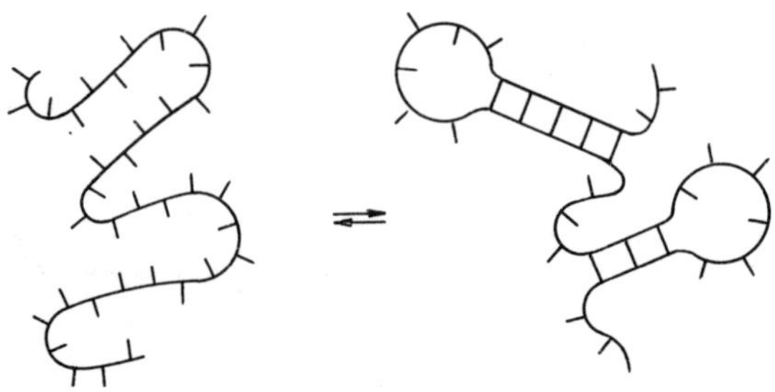

Fig. 3. Transition from single-stranded RNA to a partly double-helical structure

mRNA functioning in protein synthesis has no organized secondary structure. In the process of hybridization between RNA and DNA, the RNA will form a double helix with DNA *in vitro*. This process is often used to isolate mRNA. A modern method of isolation of mRNA of eukaryotic cells is affinity chromatography on poly(U)-sepharose*, based on the fact that mRNA of eukaryotic cells contains sequences of poly(A) of different length. As a solution of RNA molecules of different types passes through a column of poly(U)-sepharose, mRNA is specifically bound via its poly(A) segments to the column. Bound mRNA is released after elution from the column by a formamide-containing buffer. The stability of mRNA is different for different cell types as well as within any individual cells. Microbial mRNA has shorter half-lives than animal mRNA; mRNA for a constitutive enzyme has a longer half-life than that for an inducible one. The 5′-terminus of prokaryotic mRNA contains a segment which hybridizes with the 3′-terminus of 16S rRNA, facilitating binding of mRNA to the ribosome.

---

* Poly(U)-sepharose is the commercial name for small beads of agarose gel to which poly(U) chains of 30–40 nucleotides are bound covalently.

2. rRNA. Subcellular particles called ribosomes contain 50 – 60 % of rRNA. Ribosomal subunits of unequal size formed during reversible dissociation of ribosomes (see the chapter on translation of genetic information) contain RNA of different molecular weights. The smaller ribosomal subunit of *Escherichia coli* contains rRNA of molecular weight $6 \cdot 10^5$ (16S rRNA). The larger subunit yields two types of rRNA; one with $M_r = 1.2 \cdot 10^6$ (23S rRNA), the second with a molecular weight of $3.5 \cdot 10^4$, designated 5S rRNA. The 5S rRNA from *Escherichia coli* has the structure shown in Fig. 4. The RNA of both ribosomal subunits has a hairpin structure, 50 – 70 % of base residues being paired. It is assumed that the mean size of the hairpin is 25 ± 5 base residues for the smaller subunit and 35 ± 7 residues for the larger ribosomal subunit.

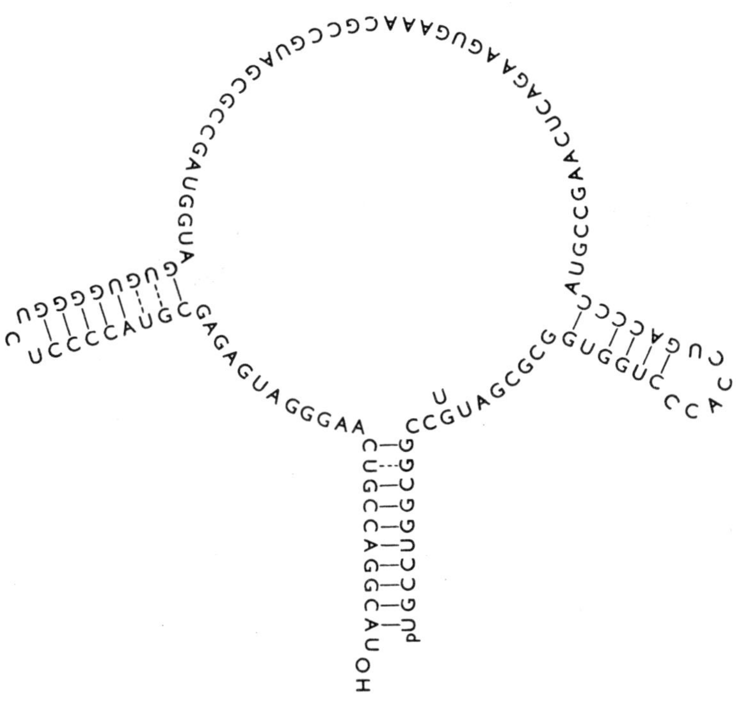

Fig. 4. One of the possible structures of 5S rRNA isolated from an *Escherichia coli* ribosome

3. tRNA. The role of tRNA in protein biosynthesis is described elsewhere. Here the structure of several tRNA types only will be mentioned. tRNA clearly differs from other RNA types in its high content of modified nucleosides (about 100 are now known). tRNAs are specific for individual amino acids and their mixture used to be called soluble RNA (sRNA). It was in tRNA that the first primary structures

of nucleic acids (i.e., sequences of nucleotides) were elucidated and secondary structures suggested. The first was that of alanine-tRNA from yeast, determined by Holley and co-workers in 1965. The various tRNAs differ only slightly from one another, as shown

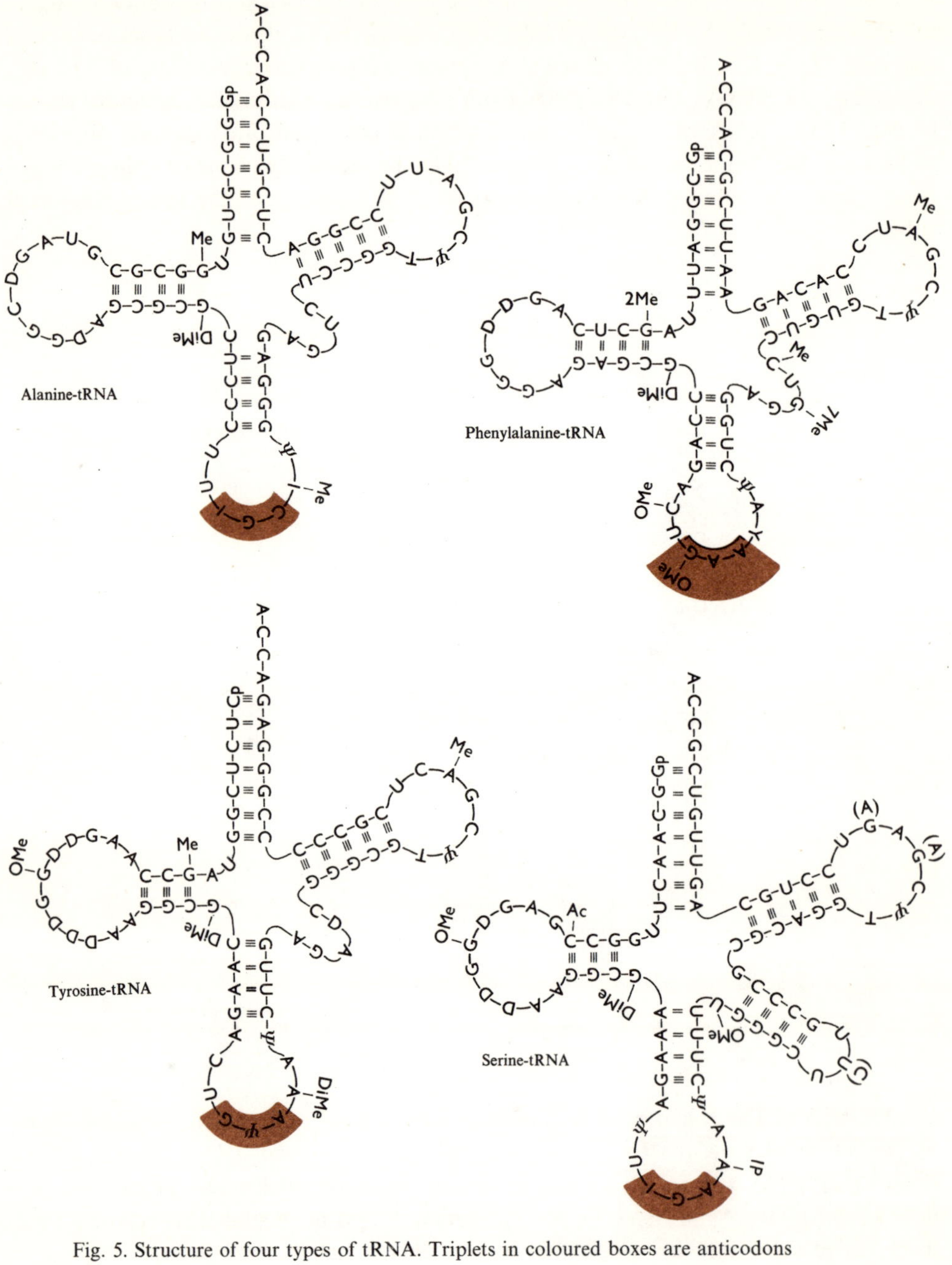

Fig. 5. Structure of four types of tRNA. Triplets in coloured boxes are anticodons

by the four structures in Fig. 5. It should be noted that more than one tRNA can exist for a given amino acid. All tRNAs have in common the tendency to form a structure reminiscent of a cloverleaf. Fig. 6 shows a generalized structure derived from the isolated and sequenced kinds of tRNA known at present (more than 60 tRNAs have now been identified and analysed). The molecular weight of tRNA is about 25 000.

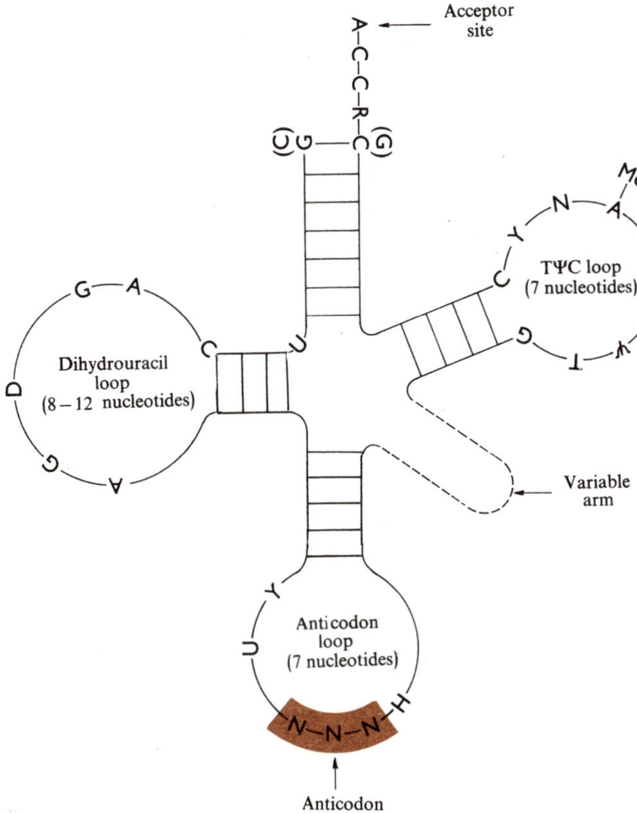

Fig. 6. Common features of all known types of tRNA
D, 5,6-Dihydrouridine; N, any ribonucleoside; R, a purine ribonucleoside; Y, a pyrimidine ribonucleoside; H, hypermodified nucleoside, such as adenosine substituted in the N^6 position with isopentenyl, tricyclic derivatives of guanosine and compounds of adenosine with amino acids (all abbreviations are listed at the end of the book)

Significant progress in research on the spatial organization of tRNA was achieved by the preparation of crystalline tRNA. X-Ray diffraction showed it to possess the cloverleaf structure, which is not planar but forms a stable L-form molecule (Fig. 7), one of the arms being terminated by the acceptor site and the other by the anticodon loop; the TΨC loop and the dihydrouracil loop are placed in the bend of the L form. This stable organization of tRNA molecules is secured by unusual types of hydrogen bonds (the so-called tertiary hydrogen bonds, in which three base residues take part in a triangular arrangement) and by van der Waals interactions of the stacking type (see p. 19).

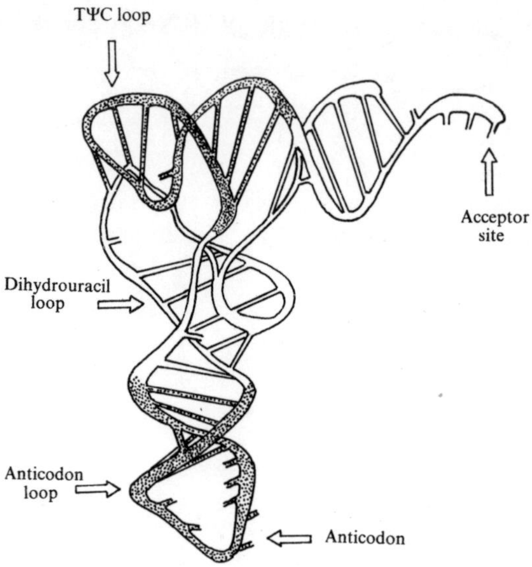

Fig. 7. Spatial arrangement of a tRNA molecule

4. Viral RNA. The smallest known RNA viruses are the so-called satellites, which can multiply only in the presence of other viruses. The molecular weight of RNA of satellite viruses is about 10^6, which is enough to code for one or two proteins only. A low molecular weight has also been reported for the RNA of bacterial viruses (phages), such as f2, R17 and M.S.2. The RNA molecular weights of other RNA viruses lie between 1.10^6 and 3.10^6. Most bacterial, plant and animal RNA viruses contain single-stranded RNA, but some of the known RNA viruses may contain an apparently double-helical structure.

Biosynthesis of Purine Precursors of Nucleic Acids

The biosynthesis of the purine components of nucleic acids is characterized by the fact that the purine backbone is not formed independently; its synthesis takes place on a primer represented by ribose 5-phosphate. The ability to synthesize the purine ring structure *de novo* is a general property of living systems. Preformed (i.e., ready-made) purines are required by only very few biological objects.

Many papers have been published on the biosynthesis of purines and it is particularly the work of Buchanan's laboratory which permits the origin of the individual atoms of the purine ring to be derived [6].

The first step in the biosynthesis of purine ribonucleotides is the phosphorylation of ribose 5-phosphate by adenosine 5′-triphosphate (ATP), which gives rise to 5-phosphoribose 1-diphosphate (PRPP) [7]. In this reaction the whole diphosphate group is transferred from ATP, as was shown by using specifically labelled radioactive ATP. The reaction is thus different from two consecutive phosphorylations effected by two molecules of ATP.

The second step in the purine backbone synthesis is the amination of PRPP by the amide nitrogen of glutamine which results in 5-phosphoribose-1-amine [8].

An alternative possibility of formation of 5-phosphoribose-1-amine was recently discovered in the enzymic reaction of ribose-5-phosphate and ammonia in the presence of ATP and magnesium salts.

[6]

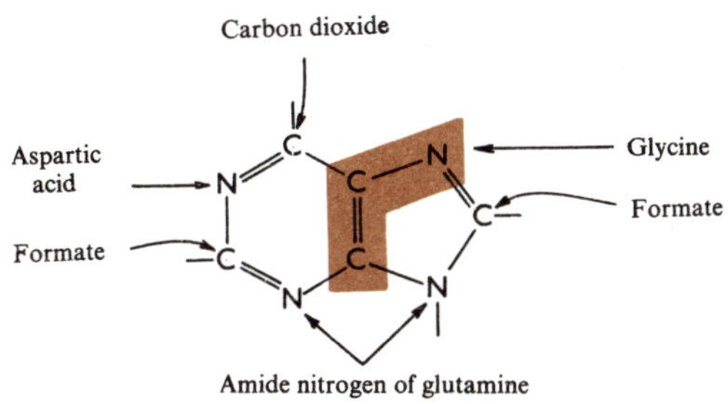

[7]

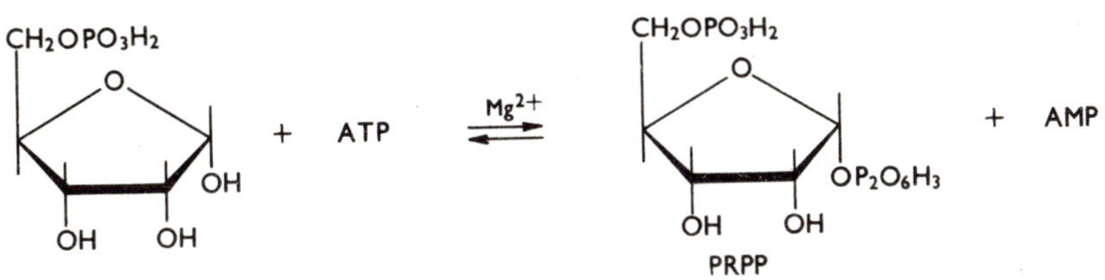

[8]

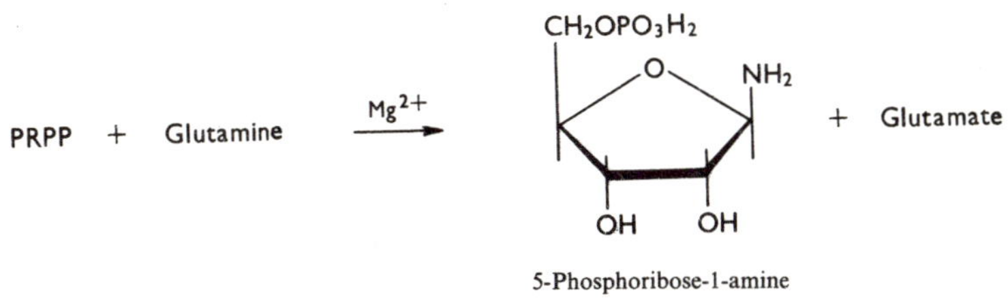

The 5-phosphoribose-1-amine has the β-configuration; the correct stereochemical orientation of the N-glycoside bond is thus determined at the very beginning of the purine ring biogenesis.

The 5-phosphoribose-1-amine is then joined by an amide bond to glycine, giving rise to glycinamide ribonucleotide [9]. The details of this enzymic reaction are not completely clear at present. The glycinamide ribonucleotide is then formylated under catalysis of N^5,N^{10}-methenyl-FH_4, where FH_4 represents tetrahydrofolate [10]. The product is then aminated by the amide group of glutamine in the presence of ATP [11].

N-Formylglycinamide ribonucleotide, which already has all the features of the imidazole ring, is then cyclized by an ATP-dependent dehydration reaction [12].

[9]

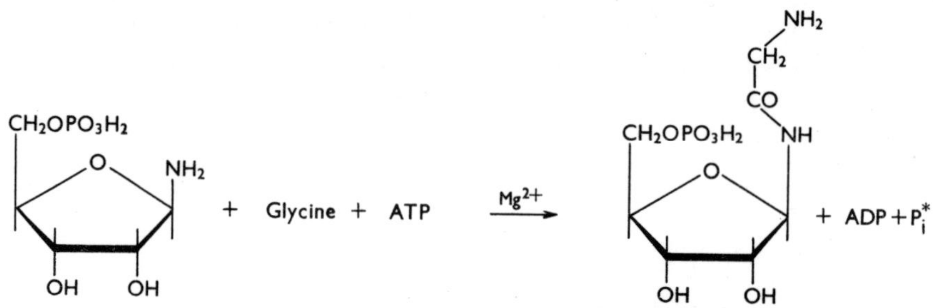

Glycinamide ribonucleotide

[10]

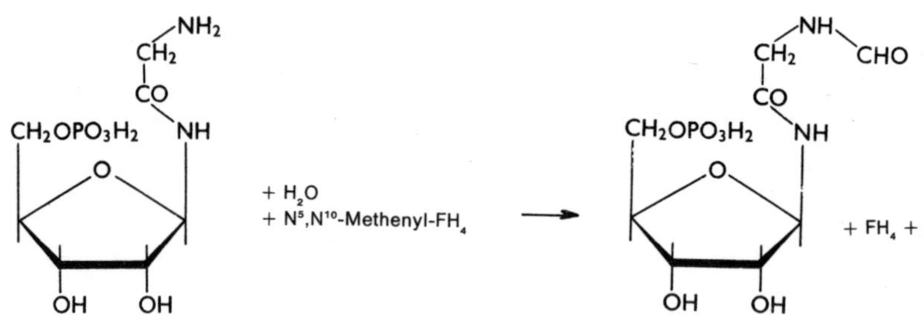

N-Formylglycinamide ribonucleotide

[11]

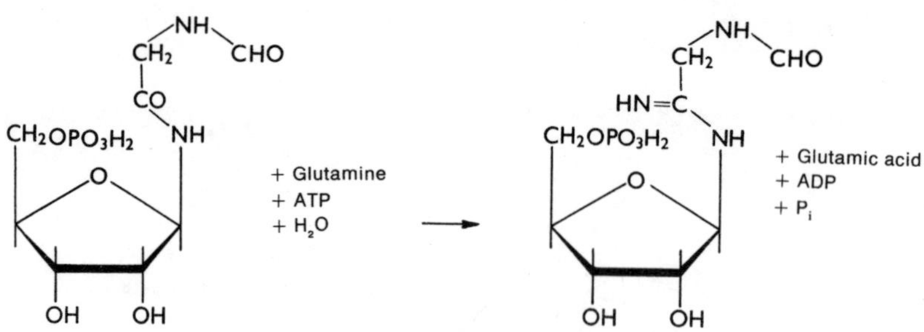

N-Formylglycinamidine ribonucleotide

* P_i = inorganic monophosphate; PP_i = inorganic diphosphate.

Under catalysis of phosphoribosylaminoimidazole carboxylase, the above ribonucleotide is carboxylated to 5-aminoimidazole-4-carboxylic acid [13]. The source of the carboxylic group for this reaction is hydrogen carbonate, which is in equilibrium with free carbon dioxide. The reaction is not known to require the presence of a coenzyme or to be coupled to an exergonic reaction.

The next step in purine nucleotide biosynthesis is the two-step formation of 5-amino-4-imidazolecarboxamide ribonucleotide. The first part of the reaction consists in a transient binding of aspartic acid to 5-aminoimidazole-4-carboxylic acid, formation of the amide bond requiring ATP [14]. This intermediate splits off fumaric acid to give rise to 5-aminoimidazole-4-carboxamide ribonucleotide [15].

[12]

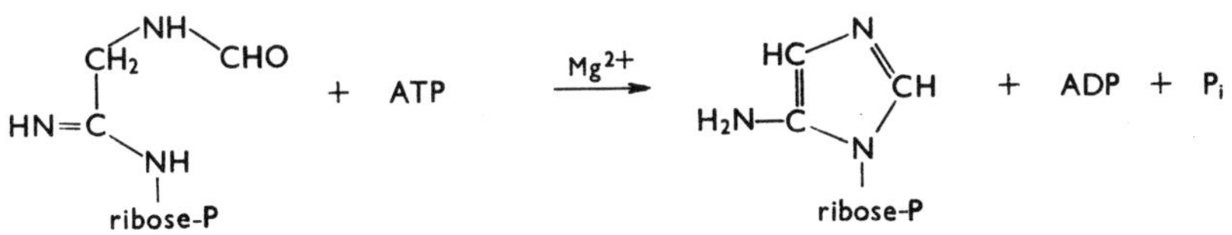

5-Aminoimidazole ribonucleotide

[13]

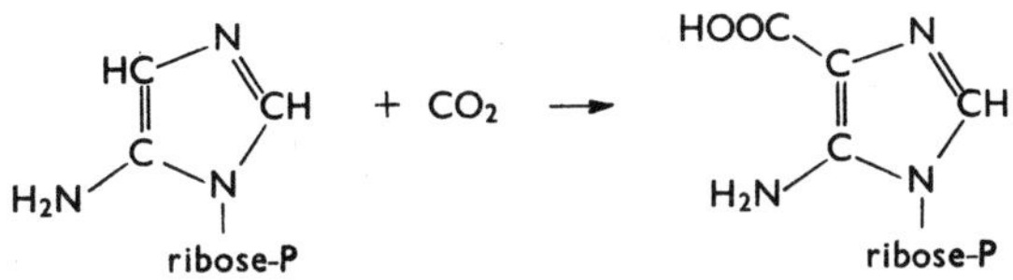

5-Aminoimidazole-4-carboxylic acid ribonucleotide

[14]

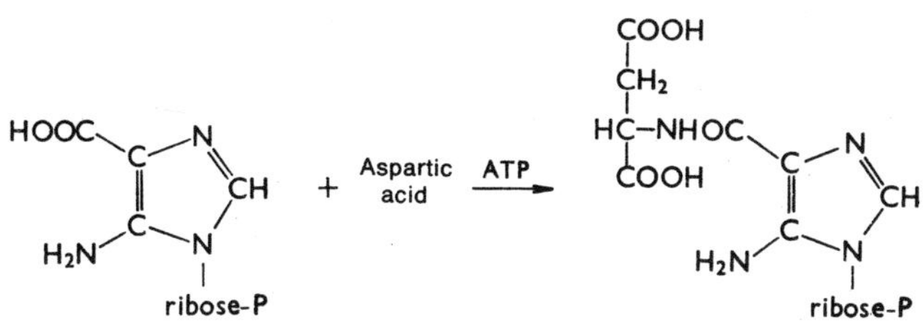

5-Aminoimidazole-4-N-succinocarboxamide ribonucleotide

[15]

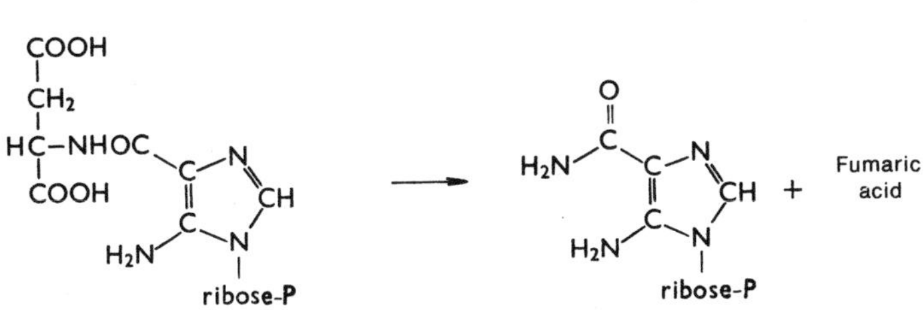

5-Aminoimidazole-4-carboxamide ribonucleotide

The last carbon atom to be introduced into the purine skeleton is provided by a formyl group with participation of N^{10}-formyltetrahydrofolic acid, which is abbreviated in equation [16] to N^{10}-formyl-FH_4.

Closure of the six-membered ring of the purine skeleton results in inosine 5′-phosphate, the base of which is called hypoxanthine [17]. Inosine 5′-phosphate is the common precursor of adenosine 5′-phosphate and guanosine 5′-phosphate, which are formed from it by the reaction sequence shown in scheme [18].

[16]

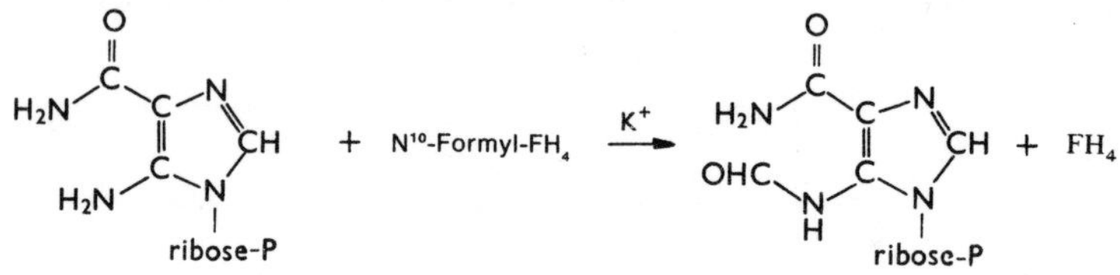

5-Formamidoimidazole-4-carboxamide ribonucleotide

[17]

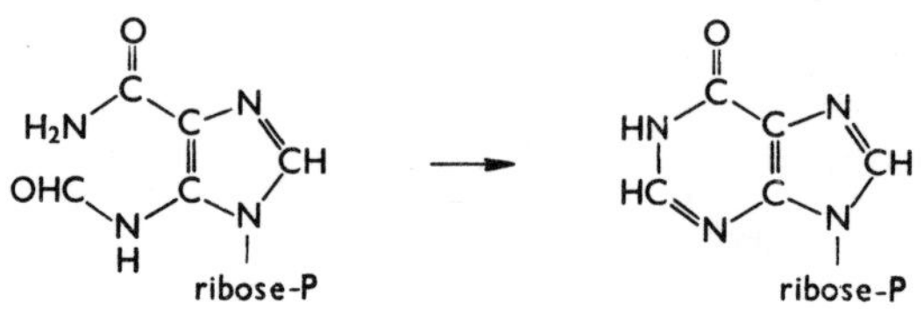

Inosine 5′-phosphate

Amination of inosine 5′-phosphate, resulting in adenosine 5′-phosphate, takes the path described above for the formation of 5-aminoimidazole-4-carboxamide ribonucleotide. The enzyme that catalyses the release of fumaric acid in both reactions is identical for both substrates. This is supported by the fact that mutants lacking the enzyme for one of the reactions are unable to perform the other reaction as well. Likewise, as the enzyme is purified, the specific activity increases for both substrates in parallel.

It is of some interest that the formation of adenylosuccinate requires GTP as coenzyme while the biosynthesis of GMP requires ATP as the coenzyme. The regulatory consequences of this sequence of controls are obvious: increased synthesis of ATP will enhance the synthesis of GTP while an increased level of GTP will have a positive effect on the synthesis of ATP.

The actual precursors of ribonucleic acids are nucleoside 5′-triphosphates which are formed from the monophosphates [19].

Kinases which are known to catalyse the above reactions are not specific for the nucleotide bases (purines and pyrimidines) and will phosphorylate even nucleotides containing some anomalous bases (antimetabolites).

[18]

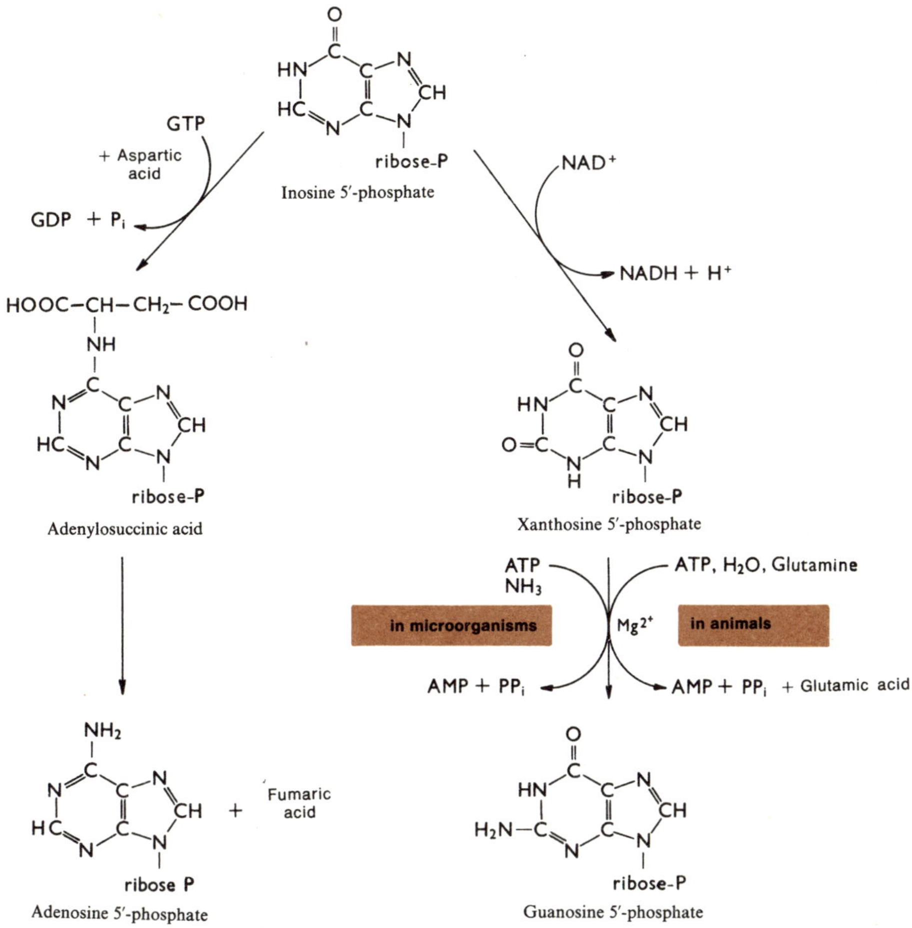

[19]

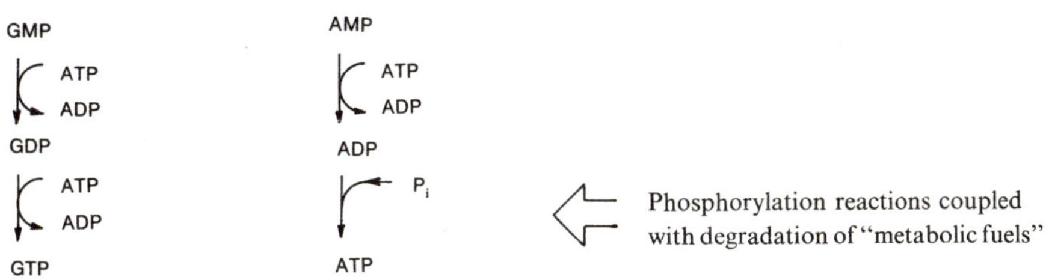

The feedback regulation of the biosynthesis of purine nucleotides is described below (p. 70).

The mutual transformations of the purine precursors of nucleic acids are summarized in scheme [20].

[20]

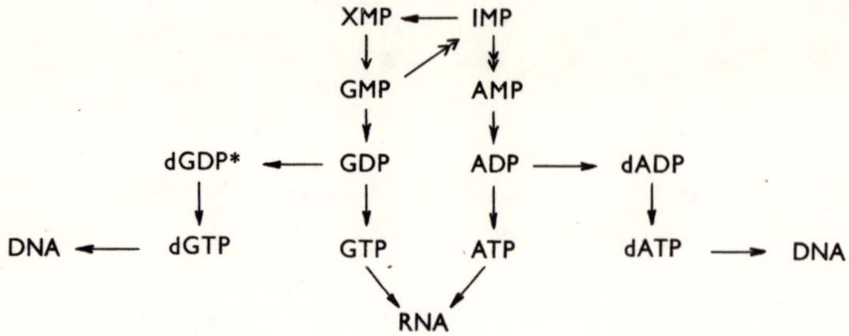

↓ Multi-step reaction

* The formation of deoxyribonucleotides is treated in the section on biosynthesis of DNA precursors.

Biosynthesis of Pyrimidine Precursors of Nucleic Acids

As shown above, the biogenesis of purine nucleotides proceeds from the very beginning at the level of ribosyl phosphates. The synthesis of pyrimidine nucleotides is different in principle as it does not require a "primer" and only a nearly complete pyrimidine base will be joined with ribosyl phosphate.

The individual atoms of the pyrimidine ring are derived from the sources shown in scheme [21].

[21]

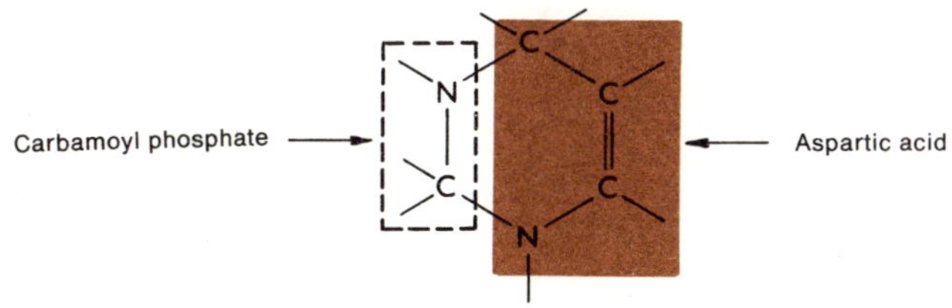

The first biosynthetic reaction of the pyrimidine ring is the formation of carbamoyl phosphate. In microorganisms it arises from carbamate,

$$NH_2COO^- + 2\,ATP \rightarrow H_2NCOOPO_3H_2 + ADP$$

and in animal systems it is formed from carbon dioxide and ammonia,

$$CO_2 + NH_3 + 2\,ATP \rightarrow H_2NCOOPO_3H_2 + 2\,ADP + P_i$$

with the participation of acetylglutamate as coenzyme, the exact role of which is unclear.

In the second step of pyrimidine synthesis the carbamoyl group is transferred to the α-amino group of aspartic acid. Aspartic acid is known to be formed either through transamination from oxaloacetic acid or directly by amination of fumaric acid.

[22]

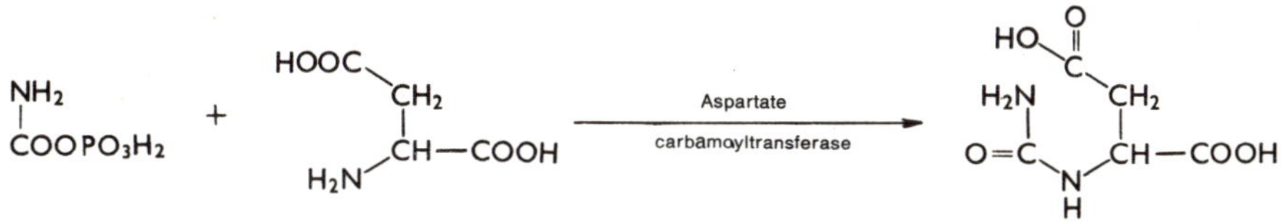

Carbamoyl phosphate Aspartic acid Ureidosuccinic acid

[23]

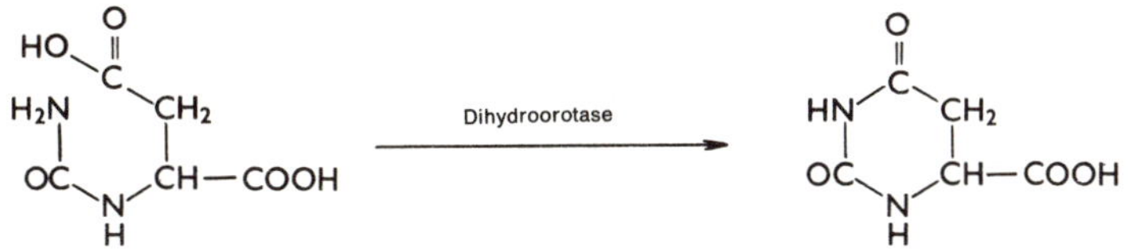

This intermediate is depicted in scheme [22]. Cyclization of ureidosuccinic acid yields dihydroorotic acid [23] and subsequent dehydrogenation gives rise to the actual precursor of pyrimidine nucleotides, orotic acid [24]. Only now will the pyrimidine ring combine with ribosyl phosphate [25] and enzymic decarboxylation of the orotidine 5′-phosphate formed will give rise to uridine 5′-phosphate [26].

[24]

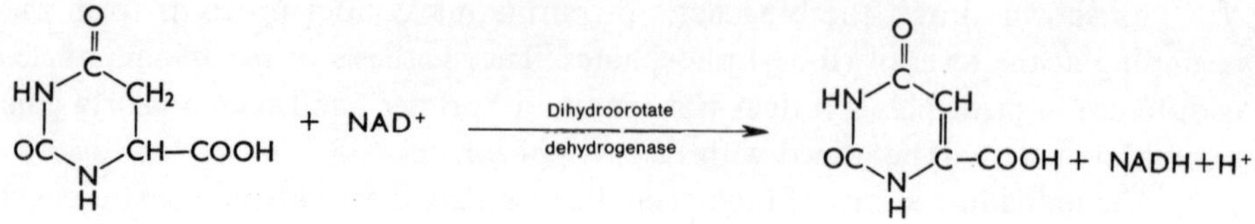

[25]

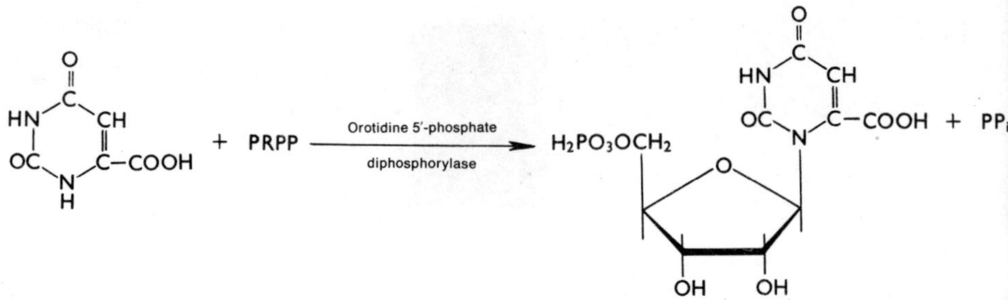

[26]

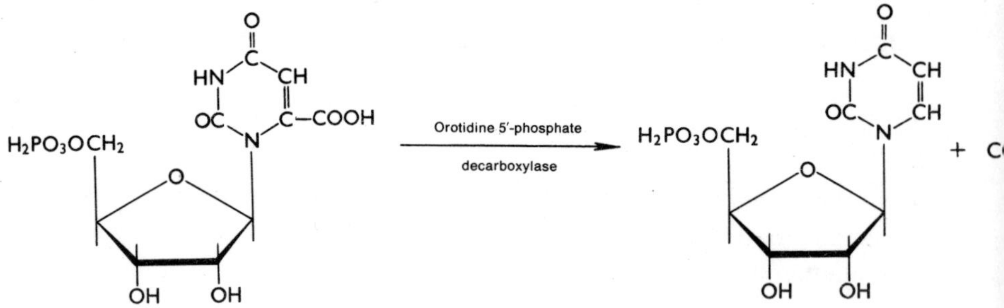

[27]

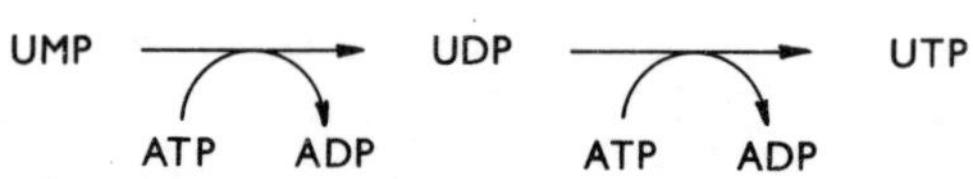

[28]

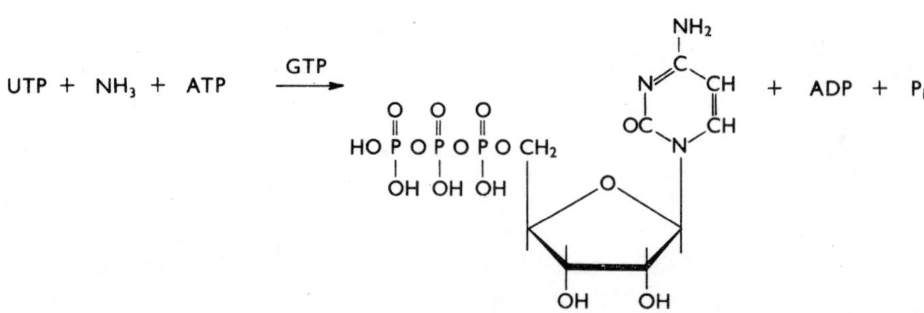

As before, the precursors of ribonucleic acids proper are nucleoside 5′-triphosphates, formed from the corresponding monophosphate in a two-step reaction catalysed by nonspecific kinases [27].

Cytidine 5′-triphosphate is formed from uridine 5′-triphosphate as shown in [28].

Biosynthesis of Precursors of DNA

The deoxyribonucleoside 5′-diphosphates are formed from the corresponding ribonucleoside 5′-diphosphates under catalysis of ribonucleoside diphosphate reductase. This enzyme, as isolated from *Escherichia coli*, is composed of two nonidentical proteins, B1 and B2; the B2 protein, in turn, is formed by two identical subunits and contains two iron atoms. The reaction is highly complex, as indicated in the scheme showing the biosynthesis of deoxycytidine 5′-diphosphate [29].

[29]

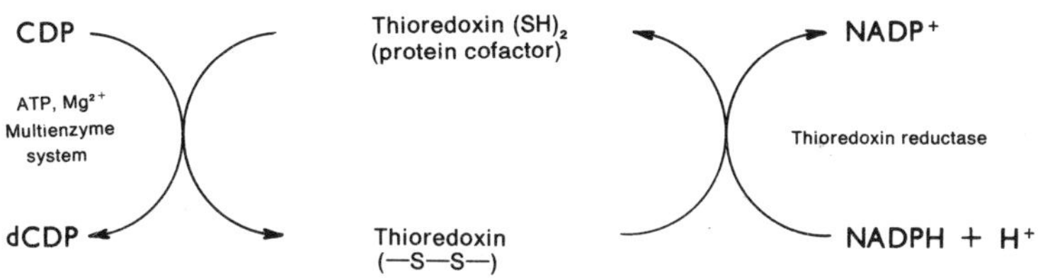

In the microorganism *Lactobacillus leichmanii* and some other prokaryotes the deoxyribonucleotides are formed from the corresponding ribonucleoside 5′-triphosphates, the reaction being catalysed by an enzyme requiring adenosylcobalamine (vitamin B_{12}) for activity. This enzyme is simpler in structure than the ribonucleoside diphosphate reductase of *Escherichia coli*.

A view has been recently advanced that in *Brevibacterium ammoniagenes* and *Micrococcus luteus* a third type of reductase is present, requiring magnesium ions for its function.

Thymidine 5′-phosphate is formed by a specific biosynthetic pathway [30].

[30]

UMP
(in some microorganisms) ↓
CMP → CDP → dCDP → dCMP → dUMP → TMP

Methylation of deoxyuridine 5′-phosphate is catalysed by thymidylate synthetase [31].

[31]

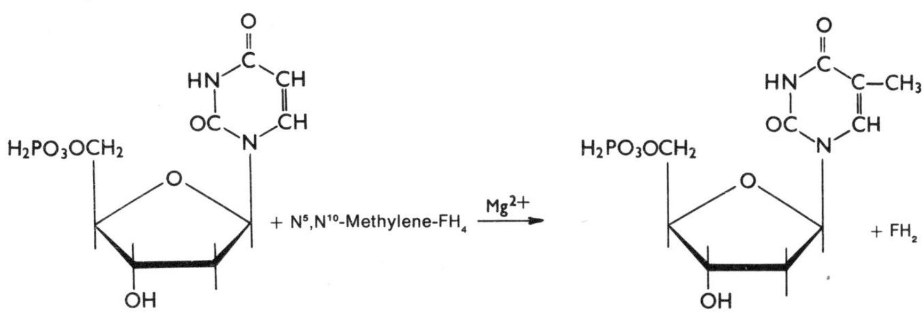

The N^5,N^{10}-methylenetetrahydrofolic acid, designated in equation [31] as N^5,N^{10}-methylene-FH_4, serves in the reaction as the carrier of one-carbon residue as well as a reducing agent. It is oxidized to dihydrofolic acid (FH_2). As the one-carbon residue is transferred to deoxycytidine 5′-phosphate, again assisted by N^5,N^{10}-methylenetetrahydrofolic acid, no reduction takes place and the pyrimidine ring then carries a hydroxymethyl instead of a methyl group.

Some biological systems, such as *Escherichia coli*, contain no dCMP deaminase; dUMP is formed here through direct reduction of UMP (an alternative reaction in the scheme shown [30]). However, dCMP deaminase can be induced here by T-even phages. These phages induce in *Escherichia coli* also deoxycytidylate hydroxymethylase, which does not substitute for thymidylate synthetase but carries out its catalytic function in parallel.

[32]

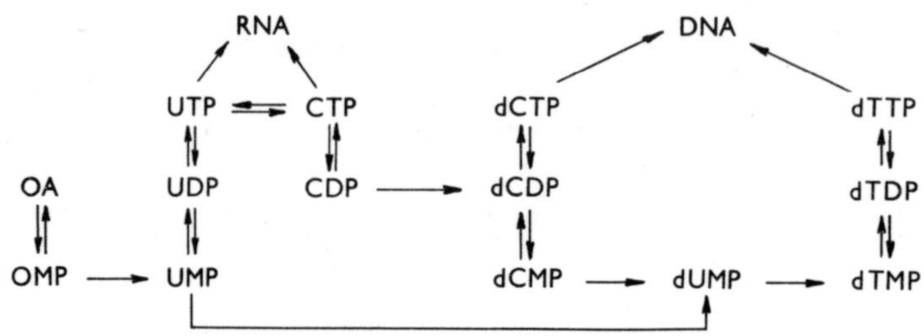

Deoxynucleoside 5′-monophosphates are then phosphorylated under catalysis of the corresponding kinases of DNA.

The metabolic pathways of the pyrimidine precursors of nucleic acids are shown in scheme [32].

Utilization of Preformed Precursors of Nucleic Acids

The biosynthetic pathways of purine and pyrimidine nucleotide formation lead to new compounds arising from simple precursors; such biogenetic pathways are designated biosynthesis *de novo*. However, many biological systems possess so-called salvage mechanisms which make it possible to utilize complete purine and pyrimidine bases or their nucleosides. The reactions involved are the following:

1. base + PRPP → nucleoside 5′-phosphate + PP_i
2. base + ribose 1-phosphate → nucleoside + P_i
3. nucleoside + ATP → nucleoside 5′-phosphate + ADP

Biosynthesis of DNA

As the genetic information stored in DNA must be transmitted by the parental generation to the filial one cells must possess a system that will ensure reproduction of DNA. If a bacterium divides into two daughter cells, each of them will under normal conditions carry its complete DNA. Hence it is necessary for the DNA to reduplicate or replicate before each division. Such a reduplication may be envisaged to take place by one of the processes shown in Fig. 8.

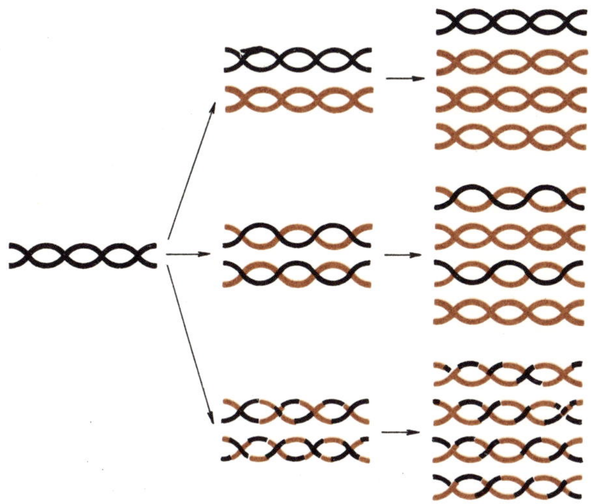

Fig. 8. Scheme of theoretically possible types of DNA replication

It was shown that *Escherichia coli* DNA replicates semi-conservatively. Cells were grown in a nutrient medium containing $^{15}NH_4Cl$ as the only source of nitrogen. After a number of cell divisions the cellular DNA will have become "heavy" owing to the content of the heavy isotope of nitrogen in its bases. This heavy DNA is readily separated from the light DNA in a gradient of caesium chloride. In fact, the DNA isolated from the ^{15}N-grown cells gave a single band after such a gradient centrifugation. If such heavy cells were then grown for just one generation time in a medium containing the normal nitrogen isotope ^{14}N, they contained DNA which, after gradient centrifugation, lay exactly between the heavy and the light DNA band positions. Analysis of DNA of further generations growing in the presence of $^{14}NH_4Cl$ showed that only three DNAs were present: the heavy one, whose content gradually decreased, the light one, whose content kept increasing, and an intermediate one, positioned half-way between the two. These findings are solid evidence for the semi-conservative type of DNA replication.

However, even with the above type of replication having been established, nothing could then be said about the mechanism itself. According to Crick and Watson, the reduplication is accompanied by partial or complete unwinding of the double helix; the exposed sections are then ready for an enzyme-catalysed synthesis of complementary new strands (Fig. 9).

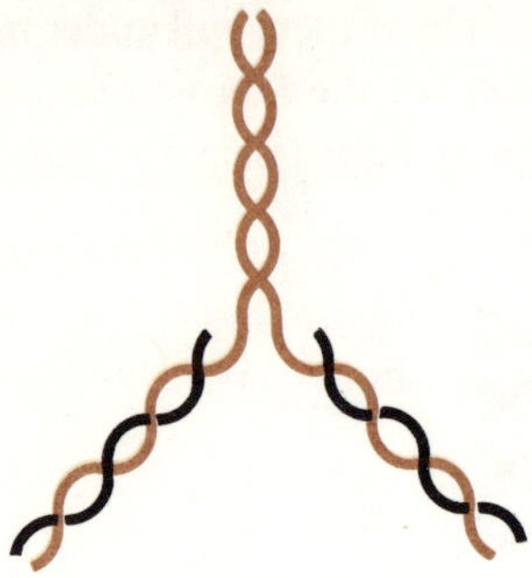

Fig. 9. Schematic outline of DNA replication

According to this theory, the identity of reduplicated DNA is guaranteed by the principle of base or nucleotide pairing; against thymidylic acid there will always be deoxyadenylic acid and vice versa, and against deoxycytidylic acid there will always be deoxyguanylic acid and vice versa.

DNA Polymerase

The enzyme catalysing the formation of polydeoxyribonucleotides was isolated by Kornberg and co-workers from *Escherichia coli* and was designated DNA polymerase or DNA synthetase (deoxynucleosidetriphosphate: DNA deoxynucleotidyltransferase). It is nowadays called DNA polymerase I and is known to catalyse the general reaction shown in [33].

$$[33]\quad \begin{matrix} n_1\ dATP \\ n_2\ dGTP \\ n_1\ dTTP \\ n_2\ dCTP \end{matrix} + DNA \xrightarrow{Mg^{2+}} DNA \left(\begin{matrix} dAMP_{n_1} \\ dGMP_{n_2} \\ dTMP_{n_1} \\ dCMP_{n_2} \end{matrix} \right. + 2(n_1 + n_2)\ PP_i$$

The reaction indicates that for DNA to be formed under catalysis by DNA polymerase all the four deoxyribonucleoside 5′-triphosphates must be present, as well as DNA as a template (it cannot be substituted by RNA) and Mg^{2+} ions. If magnesium ions are replaced with manganese, the specificity of the enzyme changes: it uses ribonucleoside 5′-triphosphates instead of the proper deoxyribonucleoside 5′-triphosphates.

Diphosphate formed during the polymerization is immediately cleaved under catalysis of pyrophosphorylase, the energy released being utilized for the biosynthesis of polynucleotides.

If the enzyme reaction is carried out *in vitro*, the maximum yield of newly synthesized DNA is about 20 times that of the original amount. The molecular weight of the polymers formed corresponds to the molecular weight of DNA introduced into the incubation mixture as a template and the content of the individual deoxyribonucleotides corresponds to the nucleotide composition of the template, as illustrated by the two examples in Table II.

Table II. Effect of DNA template on the composition of polydeoxyribonucleotides synthesized *in vitro* by DNA polymerase I

Polydeoxyribonucleotide	A	T	G	C
DNA isolated from *Escherichia coli*	1.00	0.97	0.98	1.05
Product of enzyme reaction using DNA from *Escherichia coli*	1.04	1.00	0.97	0.98
DNA isolated from *Mycobacterium phlei*	0.65	0.66	1.35	1.34
Product of enzyme reaction using DNA from *Mycobacterium phlei*	0.66	0.65	1.34	1.37

In spite of the facts concerning the reproductive activity of DNA polymerase I, some doubts were soon raised about the role of the enzyme during DNA replication *in vivo*. The physical properties of polyribonucleotides synthesized *in vitro* under catalysis of DNA polymerase are similar to those of the template, but they are not quite identical. Electron-microscopic analysis showed that the enzyme copies the DNA template not only from the beginning to the end but that it can also "return" and copy the replicated strand itself. The resulting product has then the conformation of a multiple hairpin.

In the case of partially single-stranded and partially double-stranded DNA, this DNA polymerase can repair errors and generate a normal double-helical structure.

Uncertainty about the function of DNA polymerase I in the biosynthesis of DNA molecules *in vivo* increased as a mutant of *Escherichia coli* was found that grew and replicated its DNA but that contained no DNA polymerase I activity *in vitro*. This strain was exceptionally sensitive to damage by UV light.

It is known that *Escherichia coli* cells contain at least three different DNA polymerases. The enzyme which catalyses the formation of new DNA molecules is the recently discovered DNA polymerase III. DNA polymerase I has a function in the repair mechanism and it is then understood why cells lacking the enzyme are sensitive to UV light. Nothing is known about the role of DNA polymerase II. The three enzymes differ in their molecular weight: DNA polymerase I has an M_r of 109 000, DNA polymerase II 120 000 and DNA polymerase III 180 000. They also differ in the rates at which they catalyse nucleotide polymerization, the record being held by DNA polymerase III, one molecule of which can combine 15 000 nucleotides per minute.

Even the highly purified DNA polymerases possess some 3′ → 5′-exonuclease activity. This secondary activity represents an extremely important factor for the correction of DNA structure at sites where nucleotides were paired erroneously.

DNA polymerases can attach nucleotides to already existing polynucleotide chains but they cannot initiate the synthesis of new DNA chains. It was found that the 5′-termini of DNA carry short chains of RNA and it is surmised that replication of DNA is in fact started by an RNA polymerase which catalyses the synthesis of a short segment of RNA complementary to DNA until it meets a stop signal. The RNA polymerase is then detached and the synthesized short chain of RNA serves as a primer for DNA synthesis proper, catalysed by DNA polymerase III.

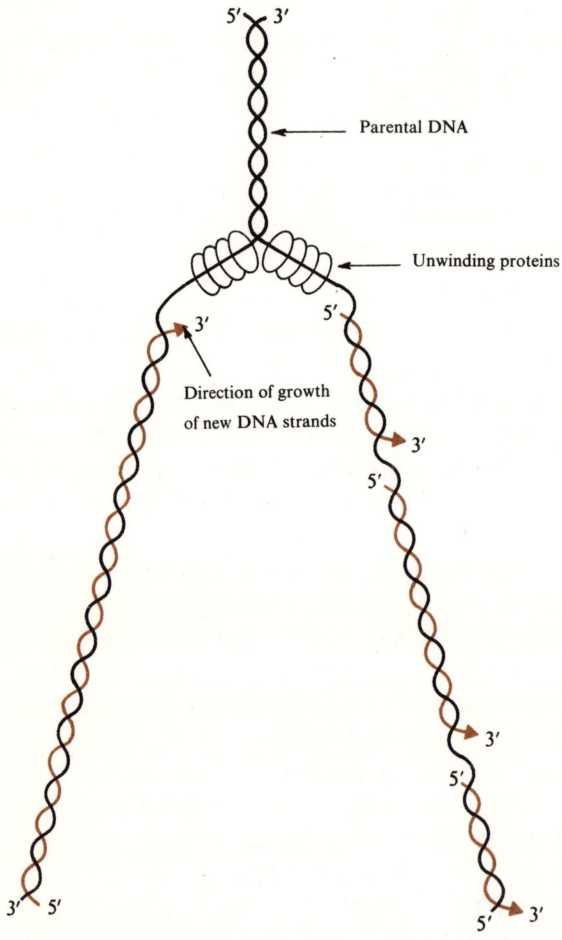

Fig. 10. Schematic illustration of a replication fork

Electron microscopy revealed that replication of a linear molecule of DNA does not start at the beginning of the strand but at about one sixth of its length. In a double-stranded DNA this gives rise to a loop (–○–) which extends both ways until it converts to a >– form, equivalent to the shorter arm of the replication fork having reached the end of the strand. It thus appears that the start of replication is marked by an unwinding of the DNA double helix whereupon the separated strands can serve as a template for replication.

Unwinding of the double-helical DNA does not take place as a consequence of attachment of new nucleotides as was first assumed. The separation of the strands is accomplished through the activity of DNA gyrase and helicase, in cooperation with the so-called unwinding proteins, which can only bind to single-strand DNA and which disrupt the double-helical organization in their vicinity (Fig. 10). A single replication fork was found to contain about 200 molecules of such unwinding proteins.

Let us now follow the fate of growing new strands in the opened section of DNA. The fact that all the known DNA polymerases catalyse the chain growth only in the direction from 5′ to 3′ led to the view that new strands on the arms of the replication fork can only grow in the opposite directions (Fig. 10). This view was in fact confirmed experimentally.

It was further found that short segments of new DNA strands are first formed, containing about 1 000 nucleotides (Okazaki fragments); these fragments are later joined together under catalysis of DNA ligase. The new strand growing in the direction of unwinding the double helix can be synthesized uninterrupted but some workers maintain that in a living cell the synthesis proceeds with intermittent breaks. It is now clear that the replication fork is a highly involved complex including other enzymes and factors.

Replication of a Circular Chromosome

A much studied example of DNA replication *in vivo* is the reduplication of the bacterial chromosome. Fig. 11 illustrates the principle of the process, which is designated bidirectional replication. The scheme also shows four sites with the so-called repair mechanisms.

Replication of the circular bacterial chromosome in cells was examined by autoradiography. Bacterial DNA was labelled with ^{3}H-containing thymine and then carefully released from cells in the presence of unlabelled bacterial DNA which served as a protective support in a micro-chamber with a membrane filter (A, Fig. 12) which permitted the passage of lysozyme and a detergent (B) toward the cells. Low-molecular-weight components released from the cells were removed by dialysis (C). The membrane filter was then punctured and the medium flowed slowly out (D). The undamaged DNA attached itself to the membrane (E) on which it was transferred to a microscope slide (F), dried and covered with a photographic emulsion. After a long exposure and deve-

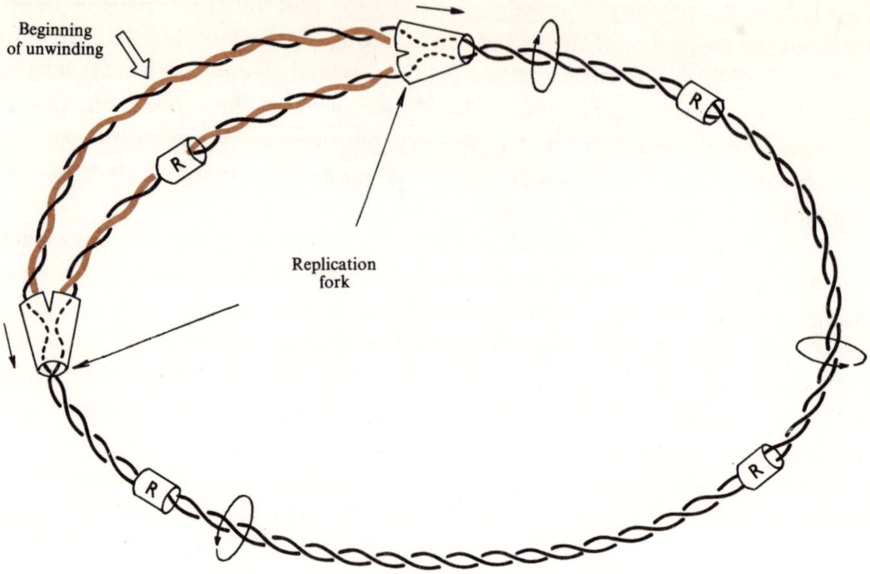

Fig. 11. Bidirectional replication of a circular chromosome. R indicates repair mechanism

lopment of the emulsion radioactive forms were revealed. A scheme of such an image (G) together with its interpretation (H) are shown in Fig. 12.

Bidirectional replication is not the only path leading to reduplication of the circular chromosome. It may also be achieved through the mechanism of a rolling circle, the principle of which is illustrated in Fig. 13. This system is used in DNA replication of many viruses and during bacterial conjugation (p. 142).

Animal cells contain four kinds of DNA polymerases. DNA polymerase α has a molecular weight of 130 000 (in growing cells it accounts for 70 – 80% of total DNA polymerase activity), DNA polymerase β has a molecular weight of 43 000 – 45 000 (5 – 15% of total activity), DNA polymerase γ of 100 000 – 120 000 (it resembles the catalytic properties of reverse transcriptase) and finally mitochondrial DNA polymerase (less than 1% of total activity). Mitochondrial DNA polymerase from rat liver has a molecular weight of 160 000. Animal cells infected with oncorna viruses (i.e., RNA viruses causing tumours) also contain reverse transcriptase, an enzyme that catalyses the synthesis of DNA on an RNA template.

DNA Topoisomerases

Topoisomerase I can split and then rejoin a single strand in double-stranded DNA; its molecular weight is 100 000. In the process the enzyme is covalently bound to the 3′ or 5′ end of the cleaved strand, via the hydroxyl of a tyrosine residue. It releases the supercoil organization of double-stranded DNA, it initiates the joining of two cyclic

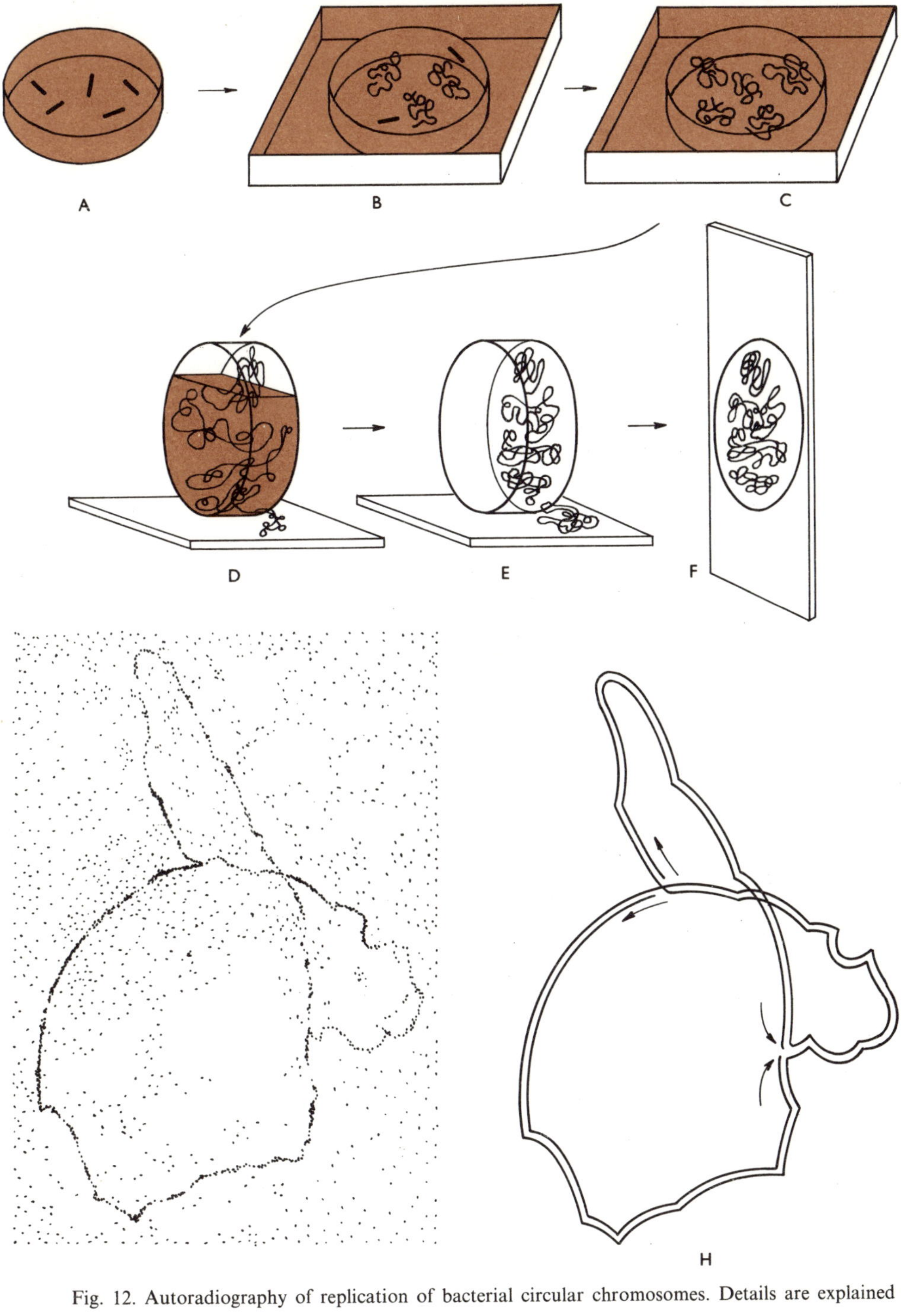

Fig. 12. Autoradiography of replication of bacterial circular chromosomes. Details are explained in the text

complementary single-stranded DNA, giving rise to a fully relaxed cyclic double-stranded DNA, and it catalyzes a number of other reactions.

Topoisomerase II contains two homologous dimers of 170 000 molecular weight. It can transiently split and rejoin double-stranded DNA. In contrast to topoisomerase I,

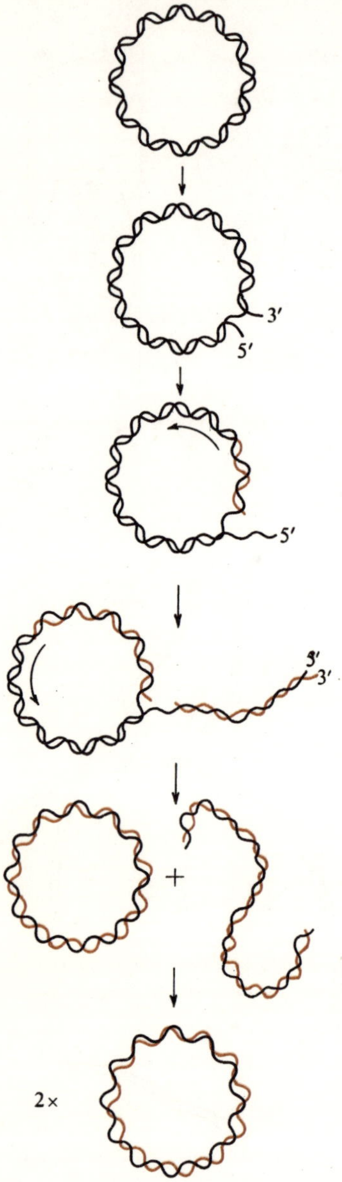

Fig. 13. Replication of a circular chromosome by the rolling circle mechanism

topoisomerase II is covalently bound via the hydroxyl group of its tyrosine always to the 5′ end of the split DNA. It removes complicated knots in DNA and it abolishes or generates the supercoil structure of DNA. The gyrase activity is included in its functions.

Both topoisomerases appear to play a control role in DNA replication.

Repair Mechanisms

It was assumed during the early days of DNA research that damage to the chemical composition of DNA either results in cell death or leads to the formation of mutants. According to recent views, however, cells possess a repair mechanism which can liquidate undesirable interference with the structure of DNA.

The repair mechanism depends on the activity of an enzyme which corrects the structure of a damaged DNA strand according to the accurate information stored in the complementary strand. Here again the principle is that of base pairing according to the Watson-Crick model.

Photoreactivation

If bacteria are irradiated with UV light of suitable intensity and wavelength only a small fraction of bacterial population will survive. However, if the irradiated cells are exposed to visible light the number of surviving cells increases several-fold. It will be seen later that UV irradiation causes thymine dimers to be formed. For clarity one may visualize the action of UV light on the DNA double helix as welding of two adjacent hooks on a zipper. A zipper after such treatment can only open up to the site of welding – by analogy, DNA replication cannot proceed beyond the site of a thymine dimer. Exposure to visible light is then analogous to cutting across the welded zipper and is achieved by a special enzyme which specializes in breaking up thymine dimers.

Dark Repair

This type of correcting mechanism was described in *Escherichia coli* mutants resistant to UV light. It was possible to prepare strains from these mutants which were again sensitive to UV light. The resistance could be transmitted by a recombination process called conjugation (p. 142). The resistant strain was found to contain no less than three genes which could be transferred by conjugation into cells sensitive to UV light. The transferred genes then control the synthesis of enzymes involved in the repair process. If the strain loses the ability to synthesize the enzymes of the repair mechanism (say, by mutation) it regains its UV sensitivity. The mechanism of dark repair is distinct from the molecular mechanism of photoreactivation. Experiments which elucidated the principle of dark repair proceeded from the study of substances released from bacteria during the repair process. Labelled thymine was incorporated into cells of both the sensitive and the resistant strains, the cells were exposed to UV light and 30 min later, the low-molecular-weight compounds were analysed. In sensitive cells the labelled thymine remained incorporated in DNA while in resistant cells some of the activity appeared in smaller fragments. This suggested that during the dark repair the damaged part of DNA is excised. Specifically, the thymine dimers were cut out from the strand rather than split as is the case with photoreactivation. The experiments with incorporated labelled thymine were repeated with the thymine analogue, 5-bromouracil. This analogue invests the fragments as well as whole DNA molecules with a greater sedimentation rate during ultracentrifugation in a density gradient; the fragments with more 5-bromouracil will move towards the bottom of the centrifuge tube. Using this technique it was possible to demonstrate that incorporation of 5-bromouracil into DNA takes place during the repair process differently from the case of normal replication and that thymine dimers are excised and replaced in the DNA strand with 5-bromouracils.

The repaired DNA has retained its ability to replicate in a semi-conservative manner, which testifies to the "vitality" of DNA after the repair process.

Action Radius of Repair Mechanisms

From what has been said about the repair mechanisms the conclusion could be drawn that they only affect the areas of radiation damage. However, it is perhaps fortunate that they will also repair damage done by nitrogen mustard which generates bonds between guanines. Moreover, the repair mechanisms will not only correct erroneous pairing but will also replace anomalous bases by normal bases. Now the repair mechanisms have been described in systems different from *Escherichia coli.* Normal human skin cells can repair damage done by excessive exposure to UV light but humans with the hereditary skin disease *xeroderma pigmentosum* lack this repair mechanism and they run the risk of a skin cancer when they expose their skin to sunlight. Thus a deficiency in the repair mechanism of skin cells is directly related here to cancerogenesis.

The repair mechanisms are responsible for the fact that DNA replication takes place with great precision. Bacterial genes in a continuous culture will replicate 10^8 times before a 50% probability arises that a single gene will be altered. On the other hand, from the point of view of evolution, such efficiently working repair mechanisms can represent a retarding factor in the development of biological species. Biological systems equipped with a flawlessly functioning repair mechanism were able to protect themselves against external influences but remained at a lower stage of development. Still, mutations controlled by repair mechanisms are only one of a number of factors of evolutionary pressure.

Biosynthesis of RNA

RNA Polymerase

The enzyme responsible for the biosynthesis of RNA is called RNA polymerase (nucleosidetriphosphate : RNA nucleotidyltransferase) and it requires for its activity the presence of all the four principal nucleoside 5′-triphosphates, of a template (DNA and RNA) and of magnesium ions [34].

$$[34]\quad \begin{matrix} n_1\,\mathrm{ATP} \\ n_2\,\mathrm{GTP} \\ n_3\,\mathrm{UTP} \\ n_4\,\mathrm{CTP} \end{matrix} \xrightarrow[\mathrm{Mg^{2+}}]{\mathrm{DNA\ (RNA)}} \left(\begin{matrix} \mathrm{AMP}_{n_1} \\ \mathrm{GMP}_{n_2} \\ \mathrm{UMP}_{n_3} \\ \mathrm{CMP}_{n_4} \end{matrix} \right. + (n_1 + n_2 + n_3 + n_4)\,\mathrm{PP_i}$$

The nucleotide composition of the nascent RNA is complementary to the template. If a copolymer of deoxyadenylic and thymidylic acid is used as template, the polyribonucleotide will contain only uridylic and adenylic acid. Template DNA and the enzymically formed polyribonucleotide were found to form specific complexes; RNA polymerase can thus catalyse the transcription of genetic information.

Using 3′-deoxyadenosine (cordycepin), it was shown that the RNA chain grows in the direction from 5′ to 3′. If this antimetabolite was added to cells it was first phosphorylated to 3′-deoxyadenosine 5′-triphosphate, which was then attached to the 3′-terminus of the RNA chain under catalysis of RNA polymerase. However, as the 3′-deoxyadenosine residue now terminating the RNA chain contains no hydroxyl group in position 3′, another nucleotide cannot be attached and the synthesis stops.

RNA polymerase has been found in a great number of biological systems and prepared in a highly purified form.

The molecular structure of RNA polymerase isolated from cells of *Escherichia coli* is complicated. The active enzyme contains five different polypeptide chains with different molecular weights (α, β, β', ω and σ with molecular weights of 39 000, 155 000, 165 000, 10 000 and 95 000, respectively). The α chain occurs twice in the RNA polymerase molecule. The enzyme has a molecular weight of about 500 000. It retains its activity even if it loses the relatively weakly attached polypeptide σ. This polypeptide has no catalytic role but (a) it is responsible for the selection of the correct strand in double-helical DNA and (b) it recognizes the start signal in the DNA molecule. Through this peptide the RNA polymerase is attached to a specific segment of DNA called the promoter, which contains the start signal. The sequence of nucleotides in many promoters is now known.

Immediately after attachment of RNA polymerase to the recognition site, the double helix of DNA locally unwinds and the RNA chain begins to be synthesized on the correct DNA strand. The recognition sites for different DNAs can be of different lengths and can contain different base sequences, but in all the cases studied the binding site for RNA polymerase contains seven nucleotides in the following sequence:

5′ TATPurATG 3′
3′ ATAPyrTAC 5′

The first nucleotide to initiate RNA synthesis in *Escherichia coli* is a purine nucleotide, usually an adenine nucleotide. The σ polypeptide is detached from the enzyme after the start of RNA synthesis. The RNA chain synthesized in the DNA template grows until the RNA polymerase reaches the terminator sequence TTTTTTA. Another type of termination signal is provided by the little understood protein called the ϱ factor.

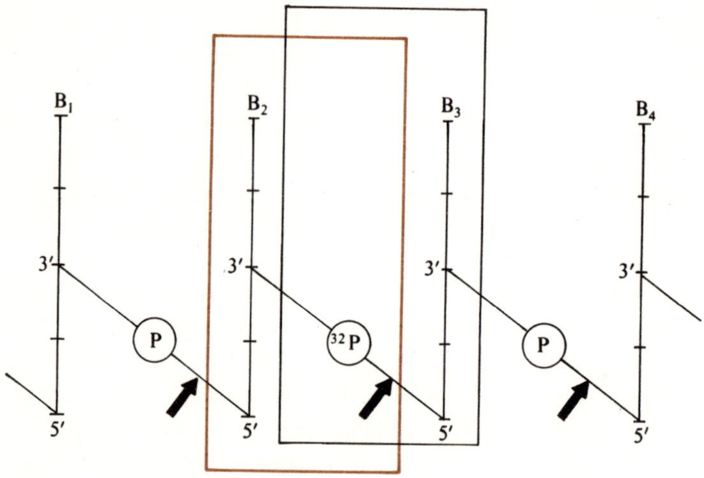

Fig. 14. Principle of the nearest neighbour method. Nucleotide in the coloured box was incorporated into the polynucleotide, nucleotide in the black box was isolated from the hydrolyzate. Arrows indicate the sites of polynucleotide cleavage during alkaline hydrolysis

The reaction catalysed by RNA polymerase may be followed by nearest neighbour analysis, as will be described below. In a ribonucleic acid containing four nucleotides a total of 16 (4^2) different dinucleotide sequences can be formed; the nearest neighbour analysis determines the frequency with which any of these sequences will appear in a given RNA preparation. The method is based on the use of incorporation of nucleoside 5′-triphosphates labelled with radioactive phosphorus in all three phosphoric residues or in that residue which is adjacent to the ribose residue. Only one of the four ribonucleoside 5′-triphosphates present in the reaction mixture is so labelled. After incorporation into the polyribonucleotide it is subjected to alkaline hydrolysis (if an analogous experiment is done with deoxyribonucleoside 5′-triphosphates and DNA-polymerase it is enzymically cleaved by a mixture of spleen and micrococcal deoxyribonuclease). This releases ribonucleoside (or deoxyribonucleoside) 3′-phosphates, which are separated by paper chromatography or electrophoresis and their radioactivity is determined. Fig. 14 indicates that the residue of labelled phosphoric acid was shifted from the introduced nucleotide to a neighbouring one, isolated after alkaline or enzymic hydrolysis.

If the reaction is carried out with one labelled nucleoside 5′-triphosphate and the three others nonlabelled in four different separate experiments the radioactivity of the nucleoside 3′-phosphates isolated after hydrolysis indicates the relative frequency with which the individual nonlabelled nucleotides were directly adjacent to the labelled one. For instance, if radioactive uridine 5′-triphosphate is used for incorporation, radioactivity is found not only in adenosine, guanosine and cytidine 3′-phosphates but also in uridine 3′-phosphate; this radioactivity of uridine 3′-phosphate then indicates the frequency of UpU in the analysed polynucleotide.

RNA polymerase can use a variety of templates: double-stranded polydeoxyribonucleotides, as well as single-stranded DNA, RNA and synthetic polyribonucleotides and oligodeoxyribonucleotides.

The composition of RNA formed under catalysis of RNA polymerase represents *in vitro* a complicated image of the template used, but *in vivo* it is truly complementary to the template used, with which it can form hybrids.

One molecule of RNA polymerase, one molecule of template DNA and the growing strand of RNA are bound during the reaction. The reaction can be stopped without releasing the reaction product and DNA will sediment together with RNA during centrifugation. We are not dealing here with a DNA/RNA hybrid as the two types of nucleic acid can be separated with dodecyl sulphate – hence they are linked through a protein, the RNA polymerase.

According to present knowledge, the nuclei of animal cells contain a greater number of different DNA-dependent RNA polymerases. Embryos of the sea urchin were found to contain three types of this enzyme; two types are present in rat liver. Polymerase I is localized in the nucleolus and polymerases II and III in nuclear plasma. α-Amanitin, a toxin from the death cap mushroom *Amanita phalloides*, inhibits polymerase II without any effect on polymerase I and III.

A number of hormones affect the biosynthesis of RNA polymerase as well as its activity. Thus, the early effect of estrogenic steroid hormones consists in a stimulation of RNA synthesis in cell nuclei of the rat uterus. The effect of hormones then depends on their firm binding to protein receptors, such as occur by the thousand in the cytoplasm of every target cell. After binding the steroid hormone, the receptor moves in complex with it towards the nucleus, where it is bound to chromatin and where it initiates previously inhibited RNA synthesis. Specific activation of RNA synthesis is probably a direct consequence of interaction of the complex steroid hormone–receptor–chromatin (i.e., not an interaction with chromosomal proteins). Histones are known to cause inhibition of RNA polymerase by their binding to DNA.

Polynucleotide Phosphorylase

Various microorganisms contain an enzyme that catalyses *in vitro* the types of reactions shown in scheme [35]. It depends on both the qualitative and the quantitative representation of individual nucleoside 5′-diphosphates in the incubation mixture in which the polymer will be formed. If a single diphosphate is present the polyribonucleotide formed is monothematic, such as polyuridylic acid, poly(U). In the presence of several types of nucleoside 5′-diphosphates, mixed polyribonucleotides are generated. The polyribonucleotide formed contains only 3′ → 5′ internucleotide bonds. If all four basic nucleotides are present artificial ribonucleic acid is formed.

[35]

$$\text{n ADP} \xrightleftharpoons{Mg^{2+}} \text{poly}(A_n) + \text{n } P_i$$

$$n_1\text{ADP} + n_2\text{UDP} \rightleftharpoons \text{poly}(A_{n_1}, U_{n_2}) + (n_1 + n_2)\ P_i$$

$$n_1\text{ADP} + n_2\text{GDP} + n_3\text{UDP} + n_4\text{CDP} \rightleftharpoons \text{poly}(A_{n_1}, G_{n_2}, U_{n_3}, C_{n_4}) + (n_1 + n_2 + n_3 + n_4)\ P_i$$

In contrast to DNA and RNA polymerases, the polynucleotide phosphorylase does not require any template for its activity but polymerization is much accelerated by the presence of a primer in the form of a short oligoribonucleotide. Polynucleotide phosphorylase thus cannot catalyse the formation of RNA with a specific sequence of bases. The biological function of this enzyme in microbial metabolism apparently consists in the phosphorolysis of RNA.

Polynucleotide phosphorylase catalyses the so-called exchange reaction [36], which is often used for assaying the enzyme activity.

[36]

$$\text{Ado-P-P} + {}^{32}P_i \xrightleftharpoons{Mg^{2+}} \text{Ado-P}^{32}\text{P} + P_i$$

Using polynucleotide phosphorylase it was possible to prepare polyribonucleotides with a most varied base content. The polymers thus formed played an extremely important role as models for interactions of polyribonucleotides and especially in unravelling the genetic code.

Methylases of Nucleic Acids

Bases of nucleic acids can be methylated at the level of finished polymers. Methylation can then be described as a post-transcriptional modification. The specific position of the methyl groups is determined by the specificity of methylation enzymes.

DNA Methylases

Escherichia coli K was found to contain an enzyme which was isolated, purified and shown to methylate cytosine in the DNA molecule to 5-methylcytosine, and adenine to 6-methylaminopurine. Methylation takes place at highly specific sites of the DNA molecule so that replacement of cytosine with methylcytosine is not accidental.

Methylation of certain bases in DNA has an important function. It protects the recognition sequences in DNA against the cell's own restriction endonucleases, which (as shown on p. 157) have become a basic tool of genetic engineering.

RNA Methylases

A number of methylases capable of methylating bases in transfer ribonucleic acids are known. The methyl group donor is S-adenosylmethionine. Five enzymes have been described in some detail. Two of them methylate guanine in position N1, one transforms uracil to thymine, another catalyses the reaction from cytosine to 5-methylcytosine and the last catalyses the formation of 2-methyladenine, 6-methyladenine and 6-dimethylaminopurine from adenine. A great deal is known about enzymes methylating ribosomal ribonucleic acids.

Specificity of Nucleic Acid Methylases

The methylation enzymes usually do not methylate homologous nucleic acids (i.e., nucleic acids isolated from the same biological system from which the methylases were obtained), as full methylation has apparently been achieved *in vivo* already. However, an enzyme isolated from a different biological species usually can add further methyl groups to heterologous DNA.

Methylation of nucleic acids plays an incontestable role in the formation of their secondary structure. This is especially significant in tRNA molecules where a definite secondary structure is essential for their function in the proteosynthetic process.

Interaction of Antimetabolites with Nucleic Acid Biosynthesis

Systematic studies with analogues of purine and pyrimidine bases of nucleic acids date back to 1950. A great number of substances were studied and many analogues with a high and largely selective biological efficiency were found or synthesized. The mechanism of action of most antimetabolites is known, some of them being mutagenic and thus representing one of the tools of experimental genetics.

An attractive area of antimetabolite study is the degree of selectivity of their effects, which occurs on several levels. A differential selectivity towards a given antimetabolite can be observed between the infective agent and the infected host; if the microorganism is more sensitive to the antimetabolite than is the host, the substance has a chance of being applied in therapy. Another level of selectivity of antimetabolite action is found between different tissues of the same organism; a special case of this selectivity is the suppression of growth of tumour tissue with antimetabolites. An important factor which can bear unfavourably on the biological effects of antimetabolites of nucleic acids is the appearance of resistance to them.

One of the first antimetabolites of nucleic acid components prepared was 6-mercaptopurine [37].

[37]

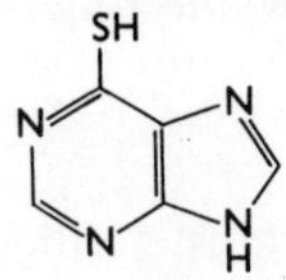

6-Mercaptopurine

In biological systems 6-mercaptopurine is transformed into 6-mercaptopurine-ribonucleoside 5′-phosphate, either directly with catalysis of pyrophosphorylase or in two consecutive steps, with catalysis of nucleoside phosphorylase and kinase. The anomalous nucleotide formed inhibits various enzymes involved in the catalysis of the last steps of biosynthesis of purine nucleotides.

6-Mercaptopurine is an antimetabolite, the action of which includes the so-called lethal synthesis. Owing to low substrate specificity of enzymes catalysing the initial steps in a metabolic sequence, antimetabolites can undergo anabolic changes. The intermediates then represent potent inhibitors of enzymes with a more expressed specificity which catalyse the later stages of the same biosynthetic pathway. Lethal synthesis is a part of action of most antimetabolites of nucleic acid components.

6-Mercaptopurine can serve to illustrate another general principle of mechanism of action of antimetabolites, viz., their polyvalent effect. Ribonucleotide formed enzymically during lethal synthesis inhibits several enzymes at once: adenylosuccinate synthetase, adenylosuccinase, inosinic acid dehydrogenase and phosphoribosyldiphosphate amidotransferase. Phosphorylation of 6-mercaptopurine riboside need not stop at the monophosphate stage. A diphosphate can be formed which inhibits polynucleotide phosphorylase, and a triphosphate, an inhibitor of nicotinamide adenylyl transferase. 6-Mercaptopurineriboside 5′-triphosphate can serve as substrate for RNA polymerase and thus the anomalous base can reach RNA and even DNA and form polymers with an altered information content. Nucleotides containing mercaptopurine can further interfere with cell metabolism via feedback inhibition of the biosynthesis of purine nucleotides, with the function of coenzyme A and finally, by competititon, with transmethylases. 6-Mercaptopurine itself is an inhibitor of xanthine oxidase.

6-Mercaptopurine is an important drug, especially for treatment of some malignant forms of the haematopoietic system. 6-Mercaptopurine and some of its derivatives (Imuran) are of interest as immunosuppressives. Such agents inhibit the synthesis of antibodies when the organism is exposed to an antigenic stimulus and are used for treatment of autoimmune diseases and after organ transplantations.

Other important purine analogues are 6-thioguanine, 8-azaguanine, 2-aminopurine and 2,6-diaminopurine [38]. These purine analogues also have a polyvalent action spectrum, are involved in lethal synthesis and are incorporated into nucleic acids.

Among pyrimidine analogues mention should first be made of 5-fluorouracil [39], which can be converted into fluorouridine 5′-mono, -di- and -triphosphate, as well as to 5-fluorodeoxyuridine 5′-phosphate. 5-Fluorouracil can be incorporated into RNA. One of the most important sites of inhibitory action of 5-fluorouracil is the inhibition of thymidylate synthetase. In uridine coenzymes (such as uridine diphosphate glucose), 5-fluorouracil can replace uracil without loss of coenzyme function. Poly-5-fluorouridylic acid acts as messenger RNA during incorporation of phenylalanine into the polypeptide. In code triplets it thus behaves in complete analogy to uracil. 5-Fluorouracil and its nucleosides are studied as potential cancerostatics.

Another inhibitor of the biosynthesis of pyrimidine precursors of nucleic acids is 6-azauracil [40].

[38]

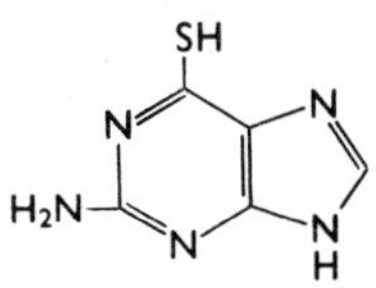

6-Thioguanine

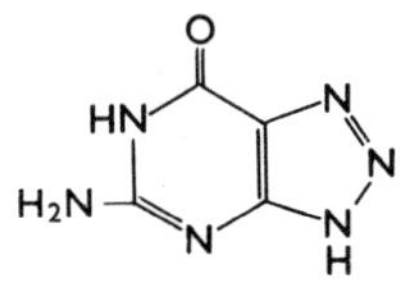

8-Azaguanine

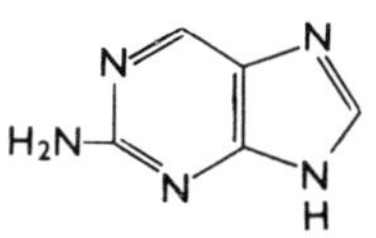

2-Aminopurine

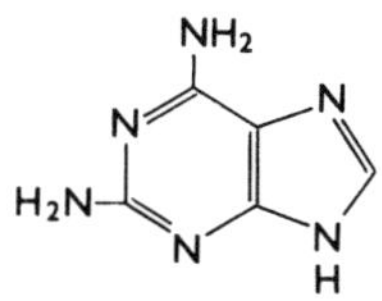

2,6-Diaminopurine

[39]

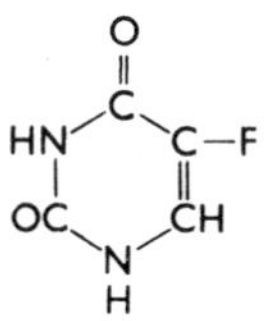

5-Fluorouracil

[40]

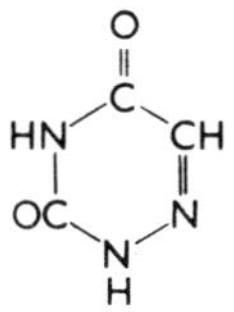

6-Azauracil

It is transformed in a variety of biological systems to 6-azauridine and 6-azauridine 5′-phosphate. This fact forms the basis of fermentative preparation of 6-azauridine. 6-Azauridine 5′-phosphate is a specific inhibitor of orotidylic acid decarboxylase. The main inhibitory action of 6-azauracil consists in blocking the formation of pyrimidine precursors of nucleic acids and of pyrimidine cofactors *de novo*.

6-Azauridine 5′-diphosphate is an inhibitor of polynucleotide phosphorylase; 6-azauridine 5′-triphosphate blocks RNA polymerase. Replacement of uracil or cytosine in the code triplets with 6-azauracil results in nonfunctional, so-called dead code triplets.

An important pyrimidine analogue in microbial genetics is 5-bromouracil [41]. Its inhibitory action is again polyvalent. It can be incorporated into DNA and this then underlies the mutagenic effects of 5-bromouracil, as will be discussed later.

Other analogues of the pyrimidine family include 5-azacytidine, which has pronounced mutagenic effects owing to its incorporation into nucleic acids, 5-iododeoxyuridine and 5-iododeoxycytidine, with some therapeutic potential, and 6-azacytidine, which resembles 6-azauridine in its inhibitory effects as well as its mechanism of action [42].

Some antimetabolites have an unchanged heterocyclic moiety but their sugar is different from that of the natural precursor. This group is represented by the synthetic antimetabolite cytosine arabinoside and the antibiotic psicofuranine [43].

[41]

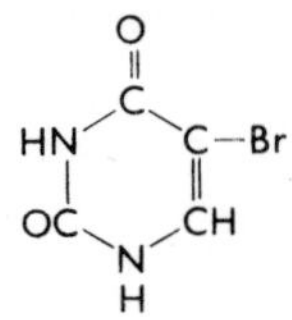

5-Bromouracil

[42]

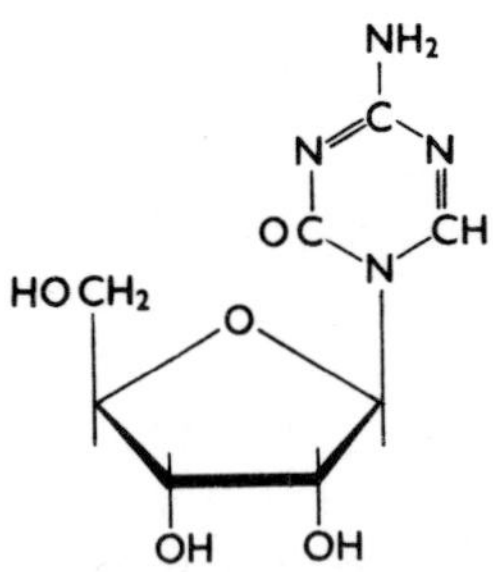

5-Azacytidine

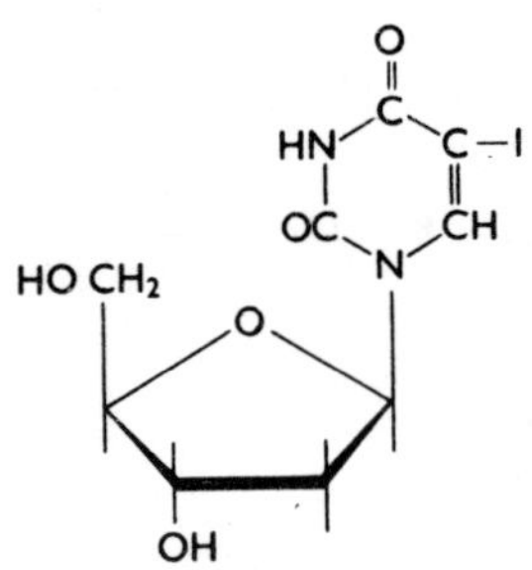

5-Iododeoxyuridine

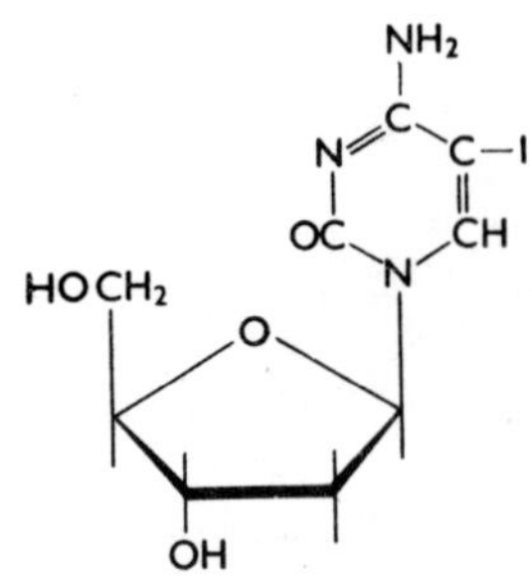

5-Iododeoxycytidine

[43]

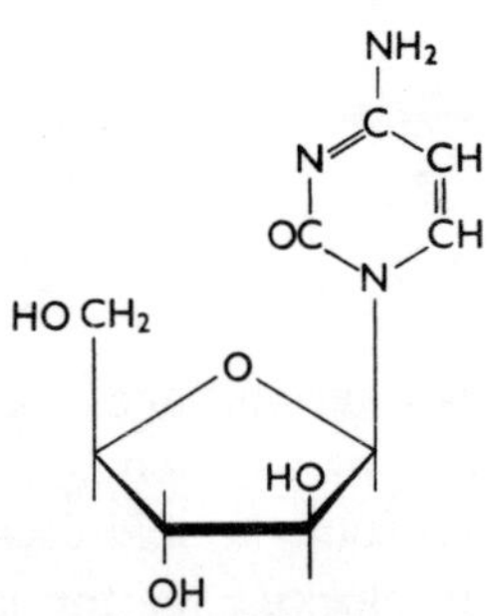

Cytosine arabinoside

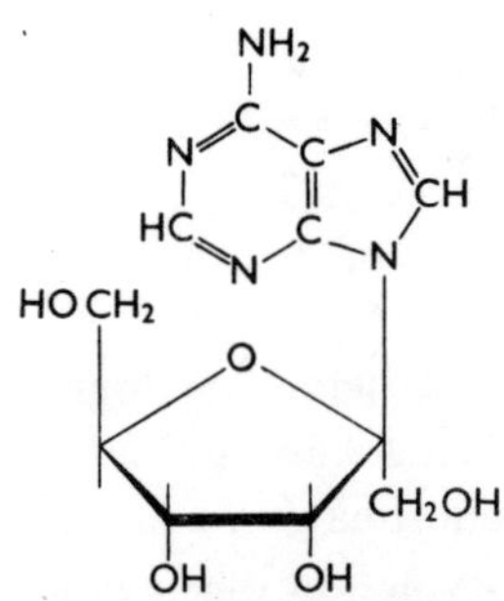

Psicofuranine

Cytosine arabinoside interferes with the reduction of cytidylic acid to deoxycytidylic acid but its incorporation into bacterial nucleic acid has not been demonstrated. Psicofuranin inhibits the final reactions of guanylic acid formation *de novo*.

The biosynthesis of nucleic acids and transfer of genetic information can also be inhibited by various antibiotics. Some of these interactions are shown in Table III.

Table III. Inhibition of nucleic acid biosynthesis and of transfer of genetic information by antibiotics and synthetic inhibitors

Area of inhibition	Inhibitor
Synthesis of DNA	aklavine
	cinerubin A and B
	daunomycin (daunorubicin)
	edein
	echinomycin
	hydroxyurea
	isoquinocycline
	mitomycin C
	nalidixic acid
	neocarcinostatin
	nitrogen mustard
	nogalamycin
	olivomycin
	phenethyl alcohol
	phleomycin
	pluramycin
	primycin
	pyrromycins
	rhodomycin
	ruticulomycin A
	sarcomycin
	streptonigrin
	tubercidin
Synthesis of RNA	aflatoxin B_1
	acridine orange
	actinomycin D
	6-azauracil
	6-azauridine
	chromomycin A_3
	cinerubin A and B
	echinomycin
	ethidium bromide
	halogenated benzimidazole ribosides
	isoquinocycline

Table III. (Cont.)

Area of inhibition	Inhibitor
Synthesis of RNA	Congo red
	cordycepin
	mithramycin
	nogalamycin
	olivomycin
	proflavin
	rifampicin
	ruticolomycin A
	streptolydigin
	streptovaricin
	thiosemicarbazone
	toyocamycin
	tubercidin
Charging of tRNA	chalcomycin
	hydroxyurea
Attachment of aminoacyl-tRNA to the ribosomes	chalcomycin
	hydroxyurea
Growth of the peptide chain	amicetin
	5-azacytidine
	blasticidin S
	carbomycin
	chloramphenicol
	cycloheximide
	erythromycin
	fusidin
	gougerotin
	lincomycin
	mycamycin B
	nucleocidin
	oleandomycin
	pactamycin
	puromycin (see Fig. 18)
	sparsomycin
	spiramycin
	streptogramin A
	streptovitacin A
	tenuazonic acid
	tetracyclines

Table III. (Cont.)

Area of inhibition	Inhibitor
Erroneous reading of genetic information	5-azacytidine gentamycin kanamycin neomycin streptomycin
Biosynthesis of ribosomes	8-azaguanine chloramphenicol erythromycin 5-fluorouracil levorphanol tetracyclines

Protein Biosynthesis

Central Dogma

Protein biosynthesis in the broad sense can be defined as the transfer of genetic information into the structure of a polypeptide. The structure which encodes the genetic information, and ribosomes on which a polypeptide is synthesized, were first assumed and then demonstrated to be linked by a mediator or adapter, RNA. This gave rise to the central dogma of molecular biology and genetics [44].

[44]

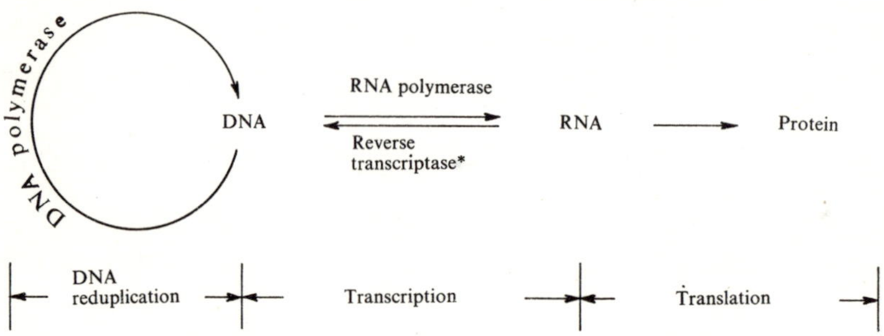

Ribonucleic acid which is transcribed on DNA is called messenger RNA (mRNA).

The preparation of a cell-free system in which the individual phases of protein synthesis could be studied was essential for understanding the whole process. Such a system was first prepared from rat liver and contained the so-called microsomal fraction with soluble enzymes and ATP. This system was able to incorporate radioactive amino acids into proteins. A similar cell-free system was later prepared from cells of *Escherichia coli* in which an "artificial mRNA" in the form of poly(U) was first used in 1961 by Nirenberg and Matthaei.

Amino Acid Activation and Formation of Aminoacyl-tRNA

Natural proteins contain 20 different amino acids. Biosynthesis of amino acids is a well established domain of biochemistry and we shall assume here that the first act of protein synthesis is the activation of an existing amino acid, which is accom-

---

* Some oncogenic viruses contain an enzyme catalysing the synthesis of DNA on a RNA template (RNA-dependent DNA polymerase or reverse transcriptase).

plished by binding the activated amino acid to a low-molecular-weight RNA. Each amino acid can use a special kind (or several special kinds) of such RNA which is called transfer RNA (tRNA).

Synthesis of tRNA molecules is achieved through transcription of specific segments of DNA localized in different parts of the chromosome. The polyribonucleotides formed, called pre-tRNA, are longer than the final tRNA and it takes the action of an endonuclease and an exonuclease to trim the pre-tRNA to the final length. The tRNA molecules then undergo further post-transcriptional modification in terms of enzyme-catalysed reactions, such as methylation, reduction and formation of pseudouridine.

The protein synthesis starts under catalysis of specific aminoacyl-tRNA synthetases (often called aminoacyl-tRNA ligases) in the following reaction:

amino $acid_A$ + ATP + $enzyme_A$ $\rightleftharpoons$ $aminoacyl_A$-AMP-$enzyme_A$ + PP_i

The complex formed in this reaction then reacts with specific tRNA as follows:

$aminoacyl_A$-AMP-$enzyme_A$ + $tRNA^A$ $\rightleftharpoons$ $aminoacyl_A$-$tRNA^A$ + AMP + $enzyme_A$.

It follows that aminoacyl-tRNA synthetases have the unique property of understanding the "language" of proteins as they recognize their correct amino acids, as well as the "language" of nucleic acids as they transfer the amino acid only to the correct specific tRNA. Aminoacyl-tRNA synthetase thus contains two specific binding sites, one for the amino acid and the other for the tRNA, plus a binding site for ATP.

The subsequent section of polypeptide chain synthesis is fully dominated by the type of tRNA. The following experiment provided the evidence. Cysteine-tRNA was charged with radioactive cysteine using cysteinyl-tRNA synthetase and then desulphurated with Raney nickel, which transformed the cysteine into alanine. An unnatural situation thus arose when tRNA for cysteine now carried alanine which should be carried by a different tRNA. Incorporation experiments then revealed that in the presence of the unnatural aminoacyl-tRNA with alanine this amino acid was incorporated at the site of cysteine. The sequence position of the amino acid was thus determined by tRNA.

Under normal conditions, each tRNA carries the correct amino acid, the only exception known being that of amino acid analogues which can also be activated and incorporated into the polypeptide chain. The specificity of aminoacyl-tRNA synthetases is thus not absolute, a case in point being that of *p*-fluorophenylalanine, which may be incorporated instead of phenylalanine.

The tRNA is now believed to contain two recognition sites, one for the mRNA (the so-called anticodon; for structure see tRNA in Fig. 6) and another for the activating enzyme.

The amino acid is attached to tRNA at the 3′-hydroxyl group of ribose in the adenosine residue of the terminal sequence pCpCpA.

Structure and Function of Ribosomes

The actual site of formation of the polypeptide chain programmed by mRNA are the ribosomes. The weight of the ribosomes may represent a substantial part of the cell dry weight; e.g., in *Escherichia coli* the fraction may be 30% and most cellular RNA is present in it. A single growing cell of *Escherichia coli* contains as many as 15 000 ribosomes. Ribosomes contain about two-thirds RNA and one-third protein by weight.

Two types of ribosomes are known in nature. Prokaryotic cells, mitochondria and chloroplasts contain smaller ribosomes sedimenting as 70S (for the 30S and 50S subunits, see Fig. 15) while eukaryotic cells contain larger ribosomes in their cytoplasm, corresponding to a sedimentation constant of 80S (with 40S and 60S subunits).

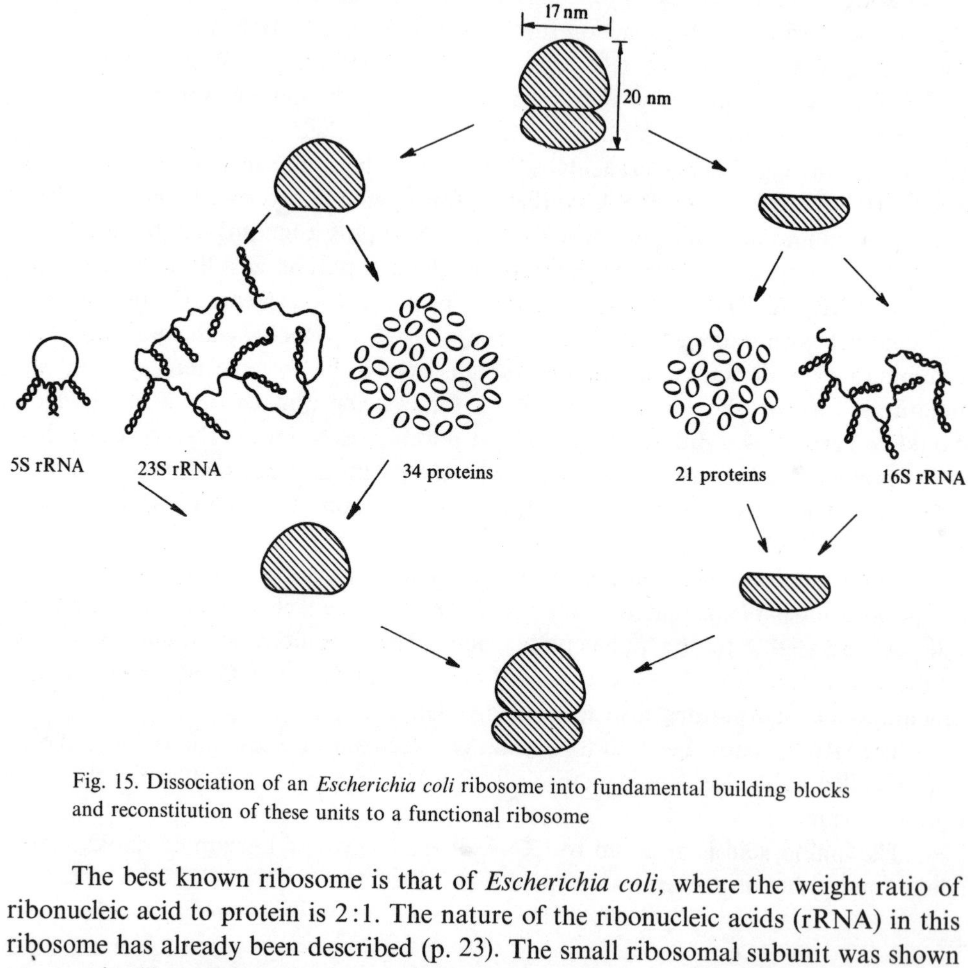

Fig. 15. Dissociation of an *Escherichia coli* ribosome into fundamental building blocks and reconstitution of these units to a functional ribosome

The best known ribosome is that of *Escherichia coli,* where the weight ratio of ribonucleic acid to protein is 2:1. The nature of the ribonucleic acids (rRNA) in this ribosome has already been described (p. 23). The small ribosomal subunit was shown to contain 21 proteins, the large ribosomal subunit 34 individual proteins, whose primary structure, localization in the ribosome and some functions are now known. Every ribosomal protein is present in a single copy. The ribosomal subunits can be broken down to their molecular components and reassembled without loss of ribosomal function (Fig. 15).

The presently accepted spatial organization of the ribosomal subunits of *Escherichia coli* is shown in Fig. 16.

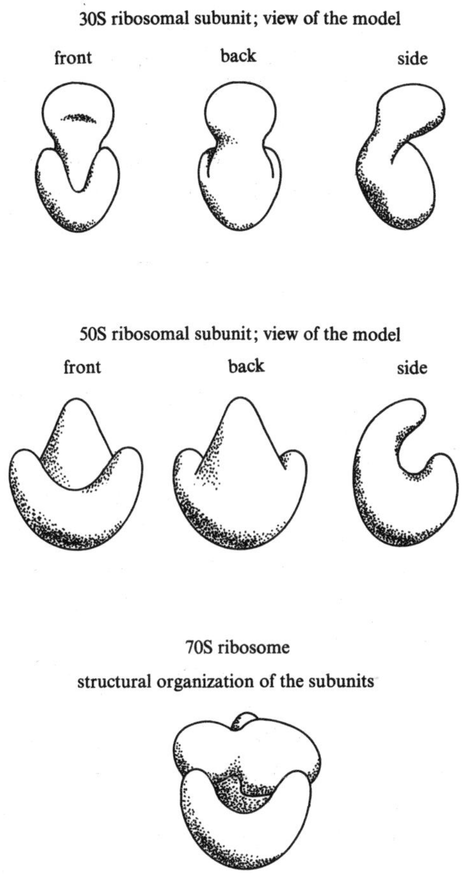

Fig. 16. Three-dimensional model of ribosomal subunits from *Escherichia coli*

Ribosomal RNA is synthesized by transcription of specific sections of DNA; such transcription first gives rise to a longer preribosomal RNA, which is then enzymically split to the individual final rRNAs (the so-called maturation).

In a eukaryotic cell nucleus transcription gives rise to the long precursor of ribosomal ribonucleic acids as well as to the so-called heterogeneous nuclear ribonucleic acid (hnRNA). This hnRNA, of very high molecular weight, is split within the nucleus enzymically to smaller fragments. It is a recent finding that within gene sequences some nonsense sections of DNA may exist. The whole DNA is first transcribed, including the nonsense segments, and their transcript is then cut out from RNA; the ends thus formed are spliced under catalysis of specific enzymes even before functional mRNA is released into the cytoplasm. The DNA segments that are spliced to form functional mRNA are called exons and the segments without apparent function are called introns. The occurrence of splicing is restricted to eukaryotic cells.

mRNA is bound to the smaller ribosomal subunit and directs the formation of a polypeptide from simple building blocks, the amino acids, imported by tRNA. As only a very small part of mRNA is in direct contact with the ribosome, the nucleic acid can attach itself to several ribosomes so that it is simultaneously translated on more than one ribosome. The complex of mRNA with several ribosomes is called polysome or polyribosome. As a single mRNA molecule is being translated simultaneously many times, a cell does not need too many mRNA molecules to achieve substantial synthesis of a given protein.

Genetic Code

Let us now examine the features of specific sequences of amino acids in a nascent polypeptide. Theory tells us that two nucleotides in a sequence in DNA could not form the code for an amino acid as, given the presence of four different deoxyribonucleotides in DNA, only 4^2, i.e. 16, code doublets are possible. If, however, the code for an amino acid contained three deoxyribonucleotides, there are 4^3, i.e. 64, possible code triplets, which is sufficient for the 20 amino acids commonly found in proteins. It will be seen that the genetic code is actually a triplet one, non-overlapping but degenerate (more than one triplet will code for the same amino acid).

The secret of the genetic code was broken in 1961 when Nirenberg and Matthaei observed that poly(U) in a cell-free system will generate polyphenylalanine. Thus the first code word was found: UUU = Phe. This led to the use of other "artificial" mRNAs for studying other codons, but a decisive role was played here by the method which applies stimulation of aminoacyl-tRNA binding to ribosomes in the presence of trinucleoside diphosphates ("triplets"). Eventually the complete code was unravelled (Table IV).

Three vicinal nucleotides in mRNA (transcribed according to the base-pairing rule of DNA) thus represent a "letter" in the genetic alphabet. This letter is then translated into an amino acid, for instance

UUU	UUA	UAU	AAU	AAA
Phe	Leu	Tyr	Asn	Lys

It was mentioned in the context of tRNA structure that a certain nucleotide triplet in those molecules may be considered as an anticodon that is compatible with the codon in mRNA. The rule of base pairing may then be slightly modified here in

Table IV. Code for translation of genetic information

First nucleotide in triplet	Second nucleotide in triplet				Third nucleotide in triplet
	U	C	A	G	
U	Phe	Ser	Tyr	Cys	U
	Phe	Ser	Tyr	Cys	C
	Leu	Ser	[a]	[a]	A
	Leu	Ser	[a]	Trp	G
C	Leu	Pro	His	Arg	U
	Leu	Pro	His	Arg	C
	Leu	Pro	Gln	Arg	A
	Leu	Pro	Gln	Arg	G
A	Ile	Thr	Asn	Ser	U
	Ile	Thr	Asn	Ser	C
	Ile	Thr	Lys	Arg	A
	Met[b]	Thr	Lys	Arg	G
G	Val	Ala	Asp	Gly	U
	Val	Ala	Asp	Gly	C
	Val	Ala	Glu	Gly	A
	Val	Ala	Glu	Gly	G

[a] Code triplets for termination of peptide chain synthesis
[b] Initiation code triplet

the sense that the last sites of the anticodon are relatively loosely defined – the so-called wobble hypothesis of Crick (Table V). This will explain why yeast $tRNA^{Phe}$ does not even contain the theoretical sequence ...ApApA... (which would be compatible with the ...UpUpU... sequence in mRNA). The anticodon triplet here is represented by the sequence ...GpApA... . On the other hand, even highly purified individual tRNAs can recognize several different codons.

Table V. Crick's wobble hypothesis

Anticodon nucleotide corresponding to the third codon position	Nucleotide of the third codon position
U	A, G
C	G
A	U
G	U, C
I	U, C, A

Initiation, Elongation and Termination of the Peptide Chain

Results of the above experiments suggested that protein synthesis proceeds linearly so that the first code word initiates the biosynthesis of the polypeptide and the other amino acids then take their turns. However, Sanger in 1964 observed a special type of amino acid bound to tRNA, viz., formylmethionine. The formyl group was bound to the amino group of methionine and it was thus clear that formylmethionine cannot form a peptide bond in which it would participate through its amino group. It followed that formylmethionine could stand only at the beginning of the polypeptide chain, as the first amino acid enters into the peptide bond through its carboxyl group. It was of some interest that the methionine amino group is formylated only after it has been bound to specific tRNA. The formylating agent here is N^{10}-formyltetrahydrofolic acid and the formylation is catalysed by a highly specific enzyme.

Table IV shows that AUG serves both as a codon for methionine and as the initiation triplet, i.e., codon for formylmethionine. Distinction between these two forms is made possible by the selective binding properties of the corresponding tRNAs: only $tRNA^{fMet}$ can bind to the initation factor whereas $tRNA^{Met}$ can only bind to the elongation factor. These factors will be treated in greater detail below.

Synthesis of the polypeptide chain requires for its initiation not only formylmethionyl-tRNA, mRNA and the two ribosomal subunits, but also three other proteins, the so-called initiation factors IF1, IF2 and IF3 which are not a permanent component of the ribosome. First, the IF3 is attached to the free 30S ribosomal subunit and the complex fMet-tRNA-IF2-GTP is formed. Both of these complexes then bind to mRNA with the aid of IF1 and then to the 50S ribosomal subunit, GTP being hydrolysed and all three initiation factors released.

It was stated above that polyphenylalanine is coded by poly (U). How can poly (U) now code for a polypeptide if it does not contain the initiation triplet AUG at its beginning? The explanation follows from the experimental conditions: if poly(U) is used as mRNA at a physiological concentration of Mg^{2+} no polypeptide is formed, but at high concentrations of Mg^{2+} artificial polyribonucleotides can govern the biosynthesis of polypeptides even without the initiation triplet.

On the basis of studying bacterial mutants producing peptide fragments, it was observed that triplets UAA, UGA and UAG represent signals for terminating the synthesis of the peptide chain. Recent sequencing analysis showed some genes to be terminated by two adjacent stop codons.

The ribosome contains two binding sites, one for binding the tRNA carrying the amino acid, the other for binding the tRNA carrying the nascent peptide. Let us call the first site the "amino acid site" and the second the "peptide site". The initiation of polypeptide synthesis on the ribosome and its first steps are shown in Fig. 17.

Continuation of the growth of the polypeptide chain on the ribosome (the so-called elongation) requires a concerted action of protein factors which are transiently bound to the ribosome. The first of these factors, elongation factor EF-Tu, stimulates in cooperation with GTP the binding of aminoacyl-tRNA to the amino acid site. Peptidyl transferase (which is a permanent component of the ribosome) then catalyses the formation of peptidyl-tRNA extended by one amino acid residue. This peptidyl-tRNA is translocated in the presence of GTP to the peptide site under catalysis of the EF-G factor. The sequence of all three reactions (binding of aminoacyl-tRNA, formation of new peptidyl-tRNA and its translocation) is called the elongation cycle;

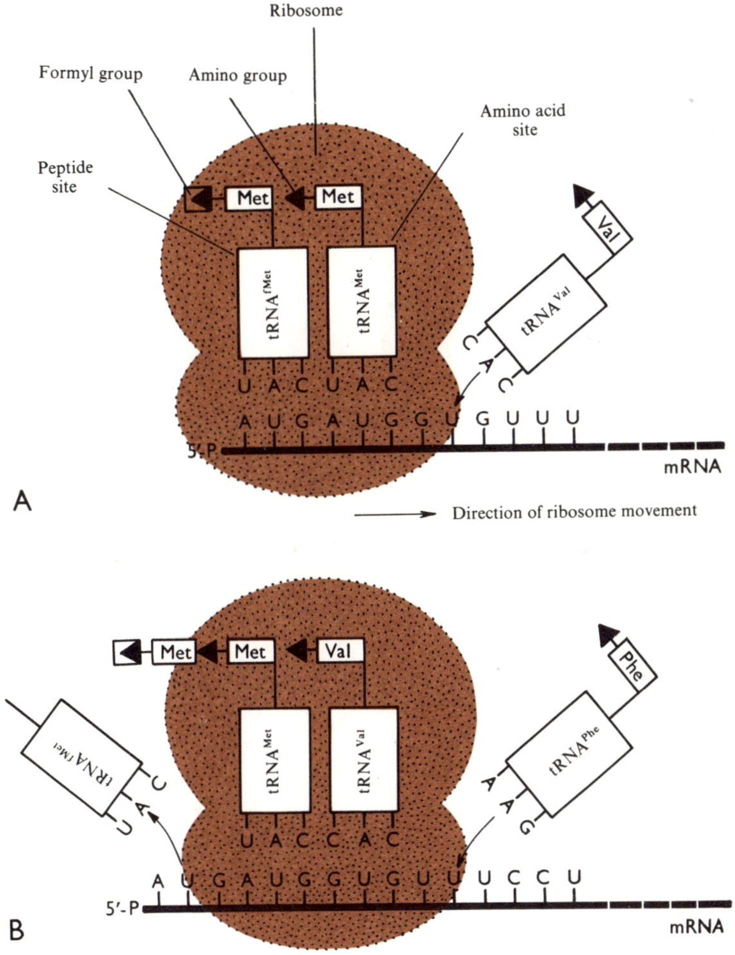

Fig. 17. Schematic representation of (A) initiation and (B) elongation of a nascent polypeptide on a ribosome

this cycle is repeated during the peptide chain synthesis as many times as there are amino acids in the resulting polypeptide. During the binding reaction and during translocation, GTP molecules are hydrolysed. The active form of the EF-Tu factor is then regenerated in conjunction with the EF-Ts factor.

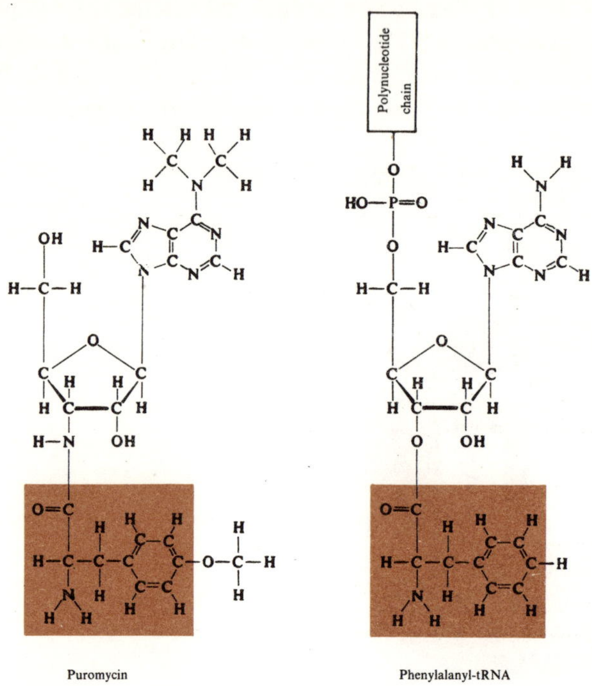

Fig. 18. Structure of puromycin and a corresponding part of the molecule of phenylalanyl-tRNA

The mechanism of formation of peptidyl-tRNA was elucidated by experiments with puromycin, an antibiotic resembling in its structure the end of the tRNA molecule which binds the amino acid. tRNA differs from puromycin by its long, partly helical polynucleotide chain which carries the anticodon. Another important difference consists in the type of bond with the amino acid in the two compounds under comparison; whereas in tRNA the amino acid is bound by an ester bond, in puromycin it is bound amidically (Fig. 18). Both compounds can bind a peptide chain through the free amino groups but bound puromycin, which has no esterically bound carboxyl group, cannot form another peptide bond and the elongation of the polypeptide is interrupted.

If it happens that an uncharged, amino acid-free tRNA molecule binds to the amino acid site, GDP will react with ATP on the ribosome to give rise to the unusual nucleotide ppGpp. The level of this nucleotide in a bacterial culture is very low under normal conditions but it increases sharply during amino acid starvation. ppGpp was found to inhibit the initiation of rRNA and tRNA chains and thus it regulates excessive formation of ribosomes and tRNA molecules.

Let us now discuss the mechanism whereby the final polypeptide chain is released from binding to the terminal tRNA. Originally it was assumed that there exist special terminating tRNAs which carry no amino acid at its 3′-hydroxyl adenosine ends; through the binding of these special tRNAs to the triplet coding for the end of polypeptide synthesis the completed polypeptide would be somehow released from the ribosomal complex. It is known now that no terminating tRNAs exist; the stop codons are "read" by specific proteins called release factors, which again require GTP for activity.

The rate of growth of the polypeptide chain is high under normal conditions: a polypeptide of M_r = 40 000 will be synthesized within 10 s. The completed polypeptide is immediately released from the ribosome and the ribosome is ready for further protein synthesis.

Isotope labelling of the large and the small subunit of the ribosome showed that an intense dissociation and association of ribosomal subunits takes place during protein synthesis. In the presence of the antibiotic sparsomycin, which blocks the elongation of the polypeptide chain, no such turnover of the ribosome occurs. Hence it appears that the separation and reassociation of the subunits are a feature accompanying protein synthesis.

The "first" amino acid in proteins of *Escherichia coli* where the problem of initiation was thoroughly studied is not always methionine but sometimes also alanine or serine. However, these amino acids have never been found formylated. Bacteria apparently contain enzymes that catalyse the cleavage of the methionine residue and thus expose other amino acids of the polypeptide chain. Hence it is possible that not only is the formyl group split off from the peptide-bound methionine but that the whole formylmethionine is cleaved as part of post-translational modifications of the polypeptide chain.

Biosynthesis of Peptide Antibiotics

The biosynthesis of peptide antibiotics (e.g., gramicidin S, tyrothricin and bacitracin) takes place in bacterial cells through a highly specialized and unique mechanism. Before incorporation of the amino acid into the antibiotic molecule it is activated in a reaction with ATP to the corresponding aminoacyladenylates. The peptide formation does not require tRNA, mRNA and ribosomes, as the amino acids are organized in their proper sequence on complexes composed of mutually linked activating enzymes. These polyenzyme complexes may have molecular weights of as much as 440 000.

Biosynthesis of a peptide antibiotic is thus a process which shares only the beginning with general protein synthesis, namely amino acid activation, all the subsequent steps being fundamentally different.

Peptide antibiotics are believed to play a role in spore differentiation in cells of sporulating bacteria. The synthesis of these antibiotics during sporulation may represent (according to Lipmann) an evolutionarily older mechanism of ordering amino acids into a specific sequence than the present-day protein synthesis. One may view this mechanism as a surviving "biochemical fossil".

Control Mechanisms

Different proteins (enzymes) are synthesized in microbial cells at different rates. Some proteins are not synthesized at all, although the required genetic information is present in the cell. Moreover, the rate of protein synthesis can differ even for a specific protein in a given cell, depending on its physiological state.

In addition to regulation of enzyme levels caused by the rate of synthesis of enzyme molecules, there exists a control mechanism influencing the actual catalytic activity of completed enzymes, e.g., feedback inhibition or feedback activation.

Feedback Control of Enzyme Activity

This is one of the principal types of metabolic rate control. If glucose labelled with radioactive carbon is fed to bacteria, the isotope will be incorporated into all amino acids of bacterial proteins. However, if the medium contains one nonradioactive amino acid, *de novo* synthesis of this amino acid will stop, as will be apparent from measuring the specific radioactivity of amino acids in a bacterial hydrolysate. Addition of this amino acid thus resulted in switching off its own biosynthesis. This switching off is very rapid, just as switching on which ensues immediately after exhaustion or removal of the added amino acid.

It is important to realize that during feedback regulation we are not dealing with suppression or stimulation of the synthesis of new molecules of an enzyme or a set of enzymes that are responsible for the formation of the given amino acid, but that only the activity of an already present enzyme is suppressed or restored.

Inhibition of a certain biosynthetic pathway by the end product is common in microorganisms and it controls the synthesis of most amino acids as well as of purine and pyrimidine precursors of nucleic acids.

The end product of a biosynthetic pathway always inhibits the enzyme catalysing the first reaction of the sequence. With branched biosynthetic pathways several enzymes catalysing the first reactions of the branches can be inhibited either individually or in concert by the end products of the branches. Feedback inhibition is shown in scheme [45].

The structure of the end product of a sequence (e.g., C_4 in scheme [45]) generally differs from that of the substrate C or of the product C_1 of the inhibited reaction. It is thus unlikely that the inhibitor would bind to the same active site as the substrate proper. Monod and co-workers proposed that the end product is bound at a site different from that of the substrate, the binding of this end product (inhibitor) to the enzyme bringing about a conformational change of the enzyme to such an extent that substrate can no longer be bound or chemically altered in the catalytic reaction. This type of inhibition is called allosteric and it differs from isosteric inhibition, where the substrate and the inhibitor are much alike and compete for the same site on the enzyme.

[45]

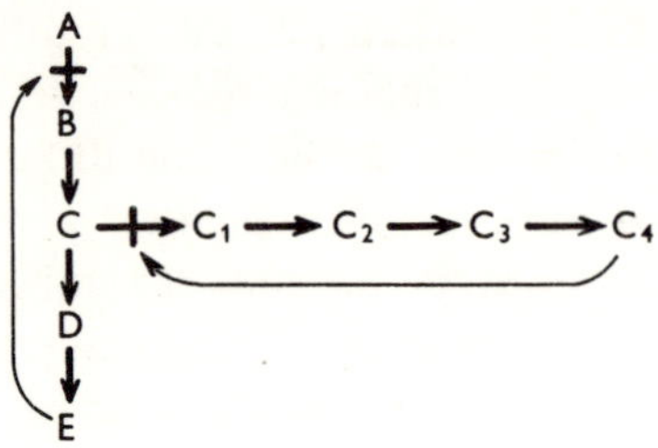

A specific example of complex feedback inhibition, such as occurs in the control of purine synthesis, is shown in Fig. 19.

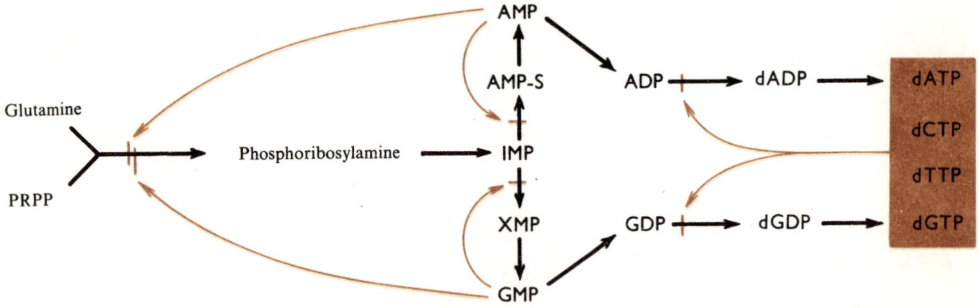

Fig. 19. Feedback inhibition controlling the biosynthesis of purine nucleotides (nucleosidic antibiotics, such as psicofuranin and decoyinin, are bound to the allosteric site of the enzyme catalysing the conversion of XMP to GMP and block its function)

Based on kinetic studies of allosteric enzymes, Jacob, Monod and Changeux developed a model which reflects most existing data well. It so happens that most allosteric enzymes are composed of two or more subunits. Each molecule can exist in conformational states I and II. If the enzyme is in state I it can bind both substrate S and activator A, but not inhibitor I. In conformational state II the enzyme can bind inhibitor I but not substrate S or activator A (Fig. 20).

As the two conformational states exist in equilibrium, binding of an activator will stabilize state I while binding of an inhibitor will stabilize state II. The enzyme can be catalytically active only in conformational state I when it can bind its substrate.

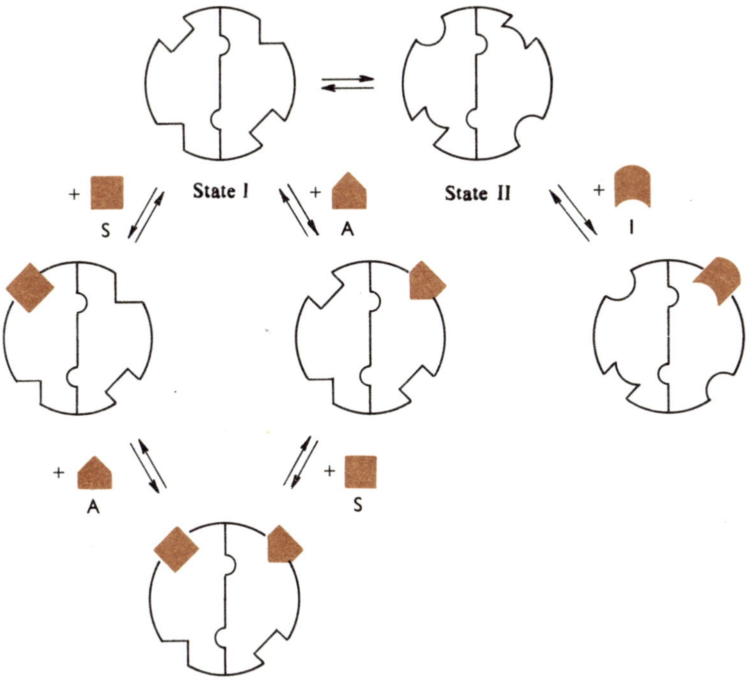

Fig. 20. Model of activation and inhibition of an allosteric enzyme composed of two subunits. S, Substrate; A, activator; I, inhibitor

Induction and Repression of Enzyme Synthesis

The level of a given enzyme or a set of enzymes can be controlled in microbial cells through the action of certain low-molecular-weight compounds. If such a compound increases enzyme synthesis in the cells, we are dealing with induction; if the enzyme synthesis is decreased, repression is involved. Both induction and repression are under the control of the genetic apparatus.

Several examples of these phenomena may help to clarify the situation. If lactose is added to a bacterial strain grown on a glucose medium a new enzyme, β-galactosidase, will soon appear in the cells. This enzyme splits the β-galactoside bond between galactose and glucose in lactose. However, the same enzyme will appear in cells on adding thiogalactoside, which is not a substrate of β-galactosidase – the so-called gratuitous inducer. If the inducer is removed from the medium, biosynthesis of β-galactosidase will stop as fast as it had begun after inducer addition. It is now clear that enzyme induction does not involve activation of a macromolecular enzyme precursor but that its biosynthesis from amino acids *de novo* sets in. Moreover, the induction of β-galactosidase formation is accompanied by that of two other enzymes, galactoside permease and transacetylase. The rate of biosynthesis is identical for all the three enzymes.

Repression is an analogous phenomenon. If histidine is added to bacteria, the biosynthesis of all enzymes of the histidine biosynthetic pathway will stop. The cause of this cessation is not an inhibition of existing enzymes (such as is the case with feedback inhibition) but a suppression of enzyme biosynthesis *de novo*. This repression can then be followed by a derepression after the histidine level drops below a certain limit.

Bacterial mutants aided in the comprehension of the genetic control of enzyme biosynthesis and the results led to the formulation of the so-called operon theory of Jacob and Monod.

The operon is a segment of DNA responsible for the formation of a given mRNA and thus represents a unit of genetic transcription. It usually contains a group of genes which regulate the synthesis of enzymes of a certain metabolic pathway. An example here is the set of genes coding for β-galactosidase, galactoside permease and transacetylase, mentioned above.

Bacterial cells contain a factor that recognizes not only substances bringing about induction or repression but, at the same time, a structure that is common to the above three enzymes. It is this structure which transmits the signal provided by lactose to initiate the synthesis of the three enzymes.

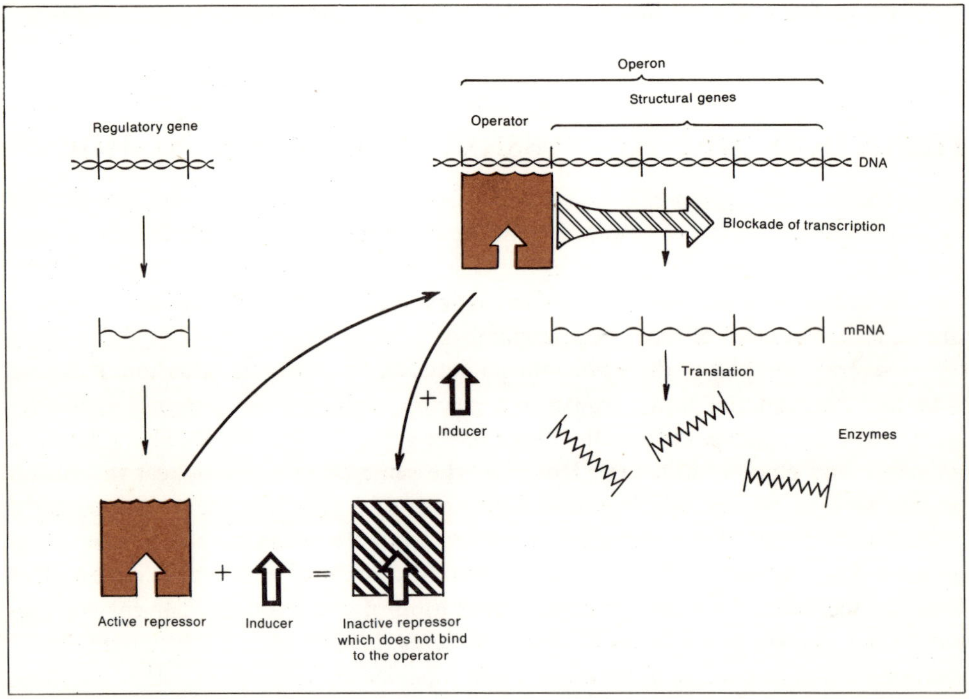

Fig. 21. Induction of enzyme synthesis

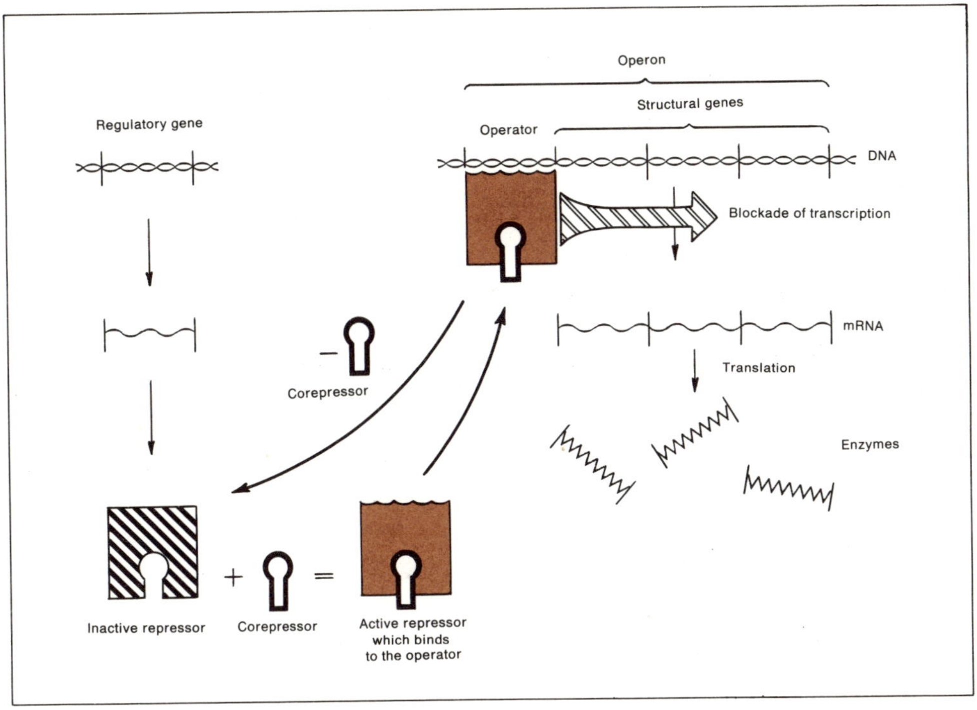

Fig. 22. Repression of enzyme synthesis

The operon theory states that under direct control of the so-called regulator gene (Fig. 21 and 22) a factor is formed which was called the repressor; such factors have been isolated and shown to be protein in nature. The repressor is inactivated by the inducer (e.g., lactose in our case) or activated by a co-repressor (e.g., histidine in another case mentioned above). Activated repressor binds to the part of DNA which is called the operator. The operator is a short segment of DNA with 21 nucleotide pairs, exhibiting a certain type of symmetry about the pair in position 11. The operator is tightly linked with its genes and forms a part of the operon. If the operator is blocked by an active repressor this means inhibition of genes coding for mRNA carrying the information for the inducible or repressible enzymes (so-called structural genes). After binding of the inducer to the repressor or after removal of the co-repressor from the repressor the operator gene is unblocked, RNA polymerase can bind to it and the corresponding structural genes of the operon can be transcribed.

The site on DNA to which RNA polymerase is bound during the start of transcription is the promoter which contains 80 nucleotide pairs (for simplicity, Figs. 21 and 22 do not show the promoters; see also p. 47).

The short half-lives of mRNA molecules for inducible enzymes were mentioned above; the mean half-life of mRNA molecules in *Escherichia coli* cells is about 2 min at 37 °C.

It had been assumed until recently that regulatory proteins represent a separate protein group with no enzyme activity of their own. Now it is certain that, e.g., in *Salmonella typhimurium*, the important enzyme glutamine synthetase can be bound to specific promoters and thus positively influence the function of two operons controlling the synthesis of four enzymes which degrade histidine with the formation of glutamate, formate and ammonia. Glutamine synthetase can function as a regulatory protein only when not adenylated; the adenylated form of the enzyme is present in a cell only with excess of assimilable nitrogen while the nonadenylated one occurs during starvation of nitrogen. By its binding to the appropriate promoters nonadenylated glutamine synthetase transmits the signal about lack of nitrogen in the cell and initiates the synthesis of enzymes breaking down histidine to the assimilable forms of nitrogen. To start this regulatory system free histidine must be present in the cell because histidine binds to the repressor of the operons involved and makes them available for binding active glutamine synthetase. This is a prudent measure on the cell's part. Even if the cell lacks nitrogen and glutamine synthetase is in its nonadenylated form, histidine must be present for the enzymes to be synthesized. If the histidine-degrading enzymes were formed even in its absence it would be highly uneconomical for the cell.

Catabolite Repression

In the presence of glucose or some other readily metabolizable sugars, the formation of some inducible enzymes is inhibited. This phenomenon is called catabolite repression. It should be noted that glucose does not interfere with the entry of the inducer (e.g., lactose) into the cell.

The promoter for the β-galactosidase operon contains some 80 nucleotide pairs. One of its regions binds RNA polymerase and another binds the complex of cyclic adenosine 3′,5′-monophosphate (cAMP) with the so-called catabolite-activating protein (CAP). This cAMP-CAP complex is essential for the specific binding of RNA polymerase.

The inhibitory effect of glucose during catabolite repression is localized in this very region and is apparently mediated by cAMP. If glucose is added one of the intermediates of its metabolism first brings about a decrease in the cAMP level in the bacterial cells. Lack of this nucleotide then precludes the formation of the cAMP-CAP complex and, without it, RNA polymerase cannot bind to the promoter to initiate transcription of mRNA.

This type of positive control ensures that glucose is utilized first, followed by other substrates, among them lactose.

Content of Nucleic Acids and Their Cellular Localization*

DNA and RNA in Microbial Cells

Before embarking on an explanation of the localization and properties of nucleic acids in bacterial cells, let us summarize the existing basic knowledge of prokaryotic cells in general.

The bacterial cell is organized according to a much simpler pattern than cells of higher biological systems. A bacterial cell is shown schematically in Fig. 23. Nearly 4 000 different types of molecules take part in the structure: one DNA, about 60 tRNAs, three rRNAs and about 1 000 mRNAs. The most numerous group are proteins, some 2 500 different molecules being present.

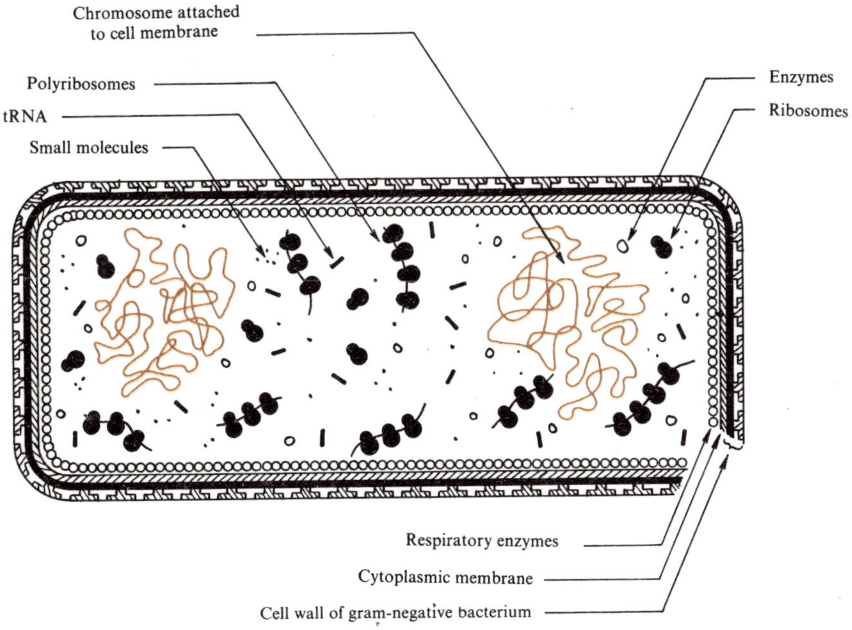

Fig. 23. Schematic cross-section of a bacterial cell

---

* According to the absolute amount of DNA in a haploid genome all organisms can be divided into four groups: I (0.001 – 0.01 pg of DNA) includes bacteria, II (0.01 – 0.1 pg of DNA) includes fungi, III (0.1 – 10 pg of DNA) includes practically all animals and many plants, IV (10 – 100 pg of DNA) includes most plants, salamanders and some fishes. Complexity of the genome thus does not necessarily reflect the degree of organismic evolution.

The bacterial cell is usually enclosed in a cell wall that protects and stabilizes the cell. Besides this, the wall is at least indirectly involved in the separation of chromosomes after division and carries some enzyme activities. It is formed by complexes of proteins, lipids and polysaccharides. The protein components contain the amino acids D-alanine, glycine, glutamic acid and α,ε-diaminopimelic acid or lysine. The polysaccharide complement of the cell wall includes N-acetylglucosamine and its ether with lactic acid, N-acetylmuramic acid. The mucopeptide or peptidoglycan built from these building blocks forms the so-called basal structure. In gram-negative bacteria the cell wall is more complex as it contains lipoproteins and lipopolysaccharides which cover the mucopeptide layer, but the cover is not completely continuous over the entire cell.

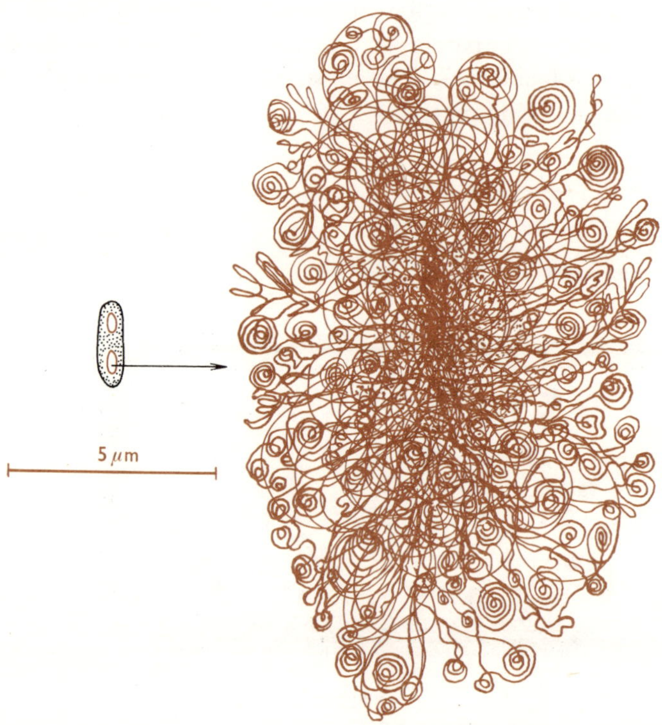

Fig. 24. Compact and extended form of a bacterial chromosome

Underneath the cell wall of the bacterial cell lies the cytoplasmic or plasma membrane composed of lipids and proteins. If the bacterial cell wall is partly (spheroplast) or completely (protoplast) removed, the cell protoplasm is contained by the plasma membrane and the cell, even if originally rod-shaped, acquires a spherical form. The plasma membrane is selectively permeable and controls the transport of substances from the medium into the cells and back. It can maintain the necessary concentrations inside the cell of both inorganic ions and most organic compounds.

The bacterial chromosome (genophore) differs in its structure and complexity from the chromosomes of higher cells. This form of nuclear equivalent of a eukaryotic cell is the nucleoid. Bacterial and phage chromosomes are single-stranded or double-stranded molecules of DNA (in some phages RNA) of different length and mass. Some bacteria and phages contain DNA, which is circular and thus closed on itself. In the bacterium *Escherichia coli* DNA takes up about 20% of the cell volume. It is packed in a very compact form, as follows from the schematic comparison of the overall length of the DNA chain and its organized form in the bacterial cell shown in Fig. 24.

Most contiguous segments of DNA are formed by genes. The individual genes containing between 600 and 1 800 nucleotide pairs are separated by only a few nucleotide pairs.

Microorganisms have a relatively higher content of nucleic acids than cells of higher organisms.

The DNA content of *Escherichia coli* represents about 1% of its wet weight. A DNA molecule represents here the entire chromosome. Under "standard" conditions a bacterial cell contains just one chromosome but owing to nonidentity of DNA replication and cell division more chromosomes may be present in a cell – as many as four. The absolute amount of DNA in a bacterial cell is $4.2 \cdot 10^{-15}$ g, the mean relative molecular weight of DNA being about $2 \cdot 10^{9}$.

The nucleotide composition of DNA is specific for every microorganism and is expressed either by the (A + T)/(C + G) ratio (see also Table I) or merely by the percentage content of G + C (Table VI).

Table VI. Representation of G + C in microbial DNA

Microorganism	% G + C
Yeast	32.4 – 45.9
Penicillium chrysogenum	50.1
Actinomycetes	73.2
Escherichia coli	49.2
Pneumococci	38.0
Clostridium perfringens	30.9

The percentage of G + C in DNA may vary even within a microbial genus. For example, Japanese microbiologists, comparing the G + C content of 119 strains of *Candida,* found it to range from 30 to 63%. In higher organisms, the interspecies differences are not so large. In mammals, for instance, the range is from 40 to 44%. However, it should be noted that DNA is heterogeneous both intra- and intermolecularly, so that the relatively less varying content of G + C in the DNA of higher organisms may be due to the presence of a much greater number of deoxynucleotides in their molecules.

Table VII. DNA and RNA content of microorganisms

Microorganism	% NA in dry solid	
	DNA	RNA*
Staphylococcus	2.8	8.8
Escherichia coli	4.2	9.7
Saccharomyces cerevisiae	0.3	3.9

* The content of RNA varies greatly depending on the physiological state of the cells

The RNA content of the microbial cell is not constant and changes in the course of culture growth. The highest RNA content is found in cells at the end of the lag phase and at the beginning of the exponential phase. The RNA content decreases sharply during the stationary phase of growth.

The ratio of DNA to RNA differs greatly for various microorganisms (Table VII).

DNA and RNA in Cells of Higher Organisms

Here again the comments on the localization of nucleic acids in the cells of fungi, algae and higher organisms (i.e., eukaryotic cells) will be preceded by a summary of information about the pertinent cell structures.

The degree of internal organization of eukaryotic cells is high. They can change their shape, shift their intracellular organelles and move as a whole. These properties are made possible by a complex cytoplasmic network of proteins, microfilaments, intermediate filaments and microtubules, which together form the cytoskeleton.

A schematic representation of a generalized eukaryotic cell is shown in Fig. 25. Such an idealized cell contains systems of membranes, the outer membrane being called plasma membrane. In most plant and some animal cells the plasma membrane is enclosed in a wall or capsule. The most conspicuous intracellular organelle is the nucleus, separated from the cytoplasm by a double membranous envelope. The cell nucleus contains DNA, RNA, proteins, lipids and inorganic solutes. It also includes a number of enzymes and metabolic intermediates. The nuclear proteins are either basic (histones*, possibly protamines) or acidic (acid or nonhistone proteins).

In a certain developmental phase of the cell the nuclear material, chromatin, is organized into filamentous and well organized structures called chromosomes.

* There are five principal types of histones that are well characterized; their molecular weights and complete amino acid sequences are known.

The dry matter of chromosomes, isolated at a certain stage of nuclear division (metaphase), was found to contain 12–20% of DNA, 10–15% of RNA and 65–75% of proteins (both histones and nonhistone proteins).

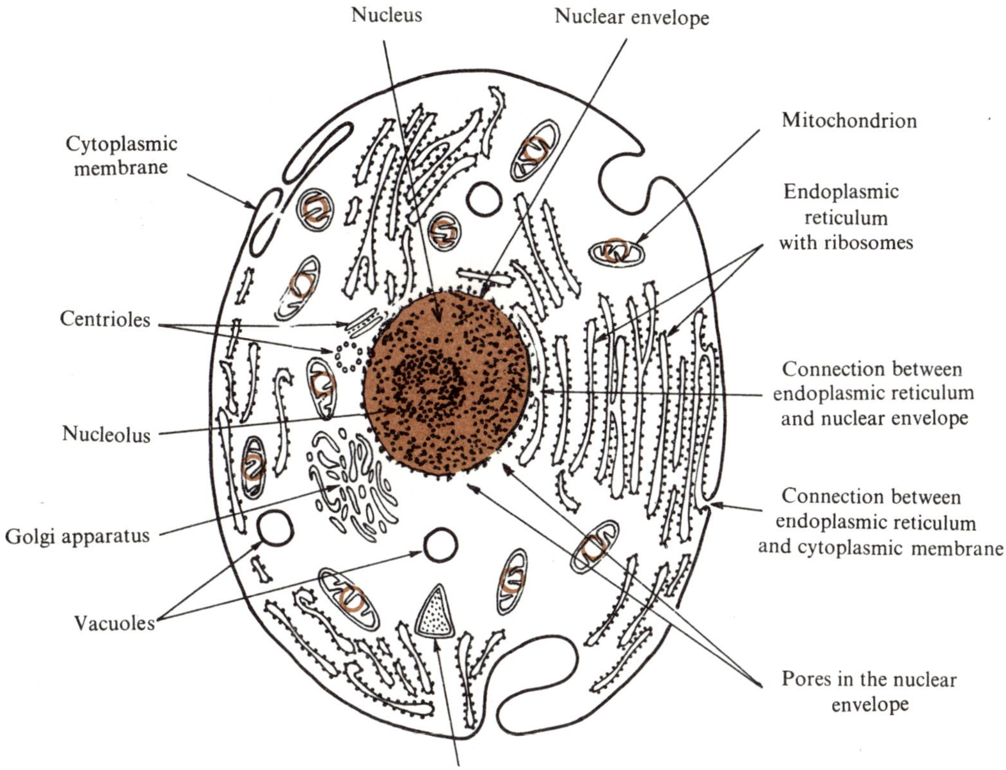

Fig. 25. Model of a cell of higher organisms (DNA is shown in colour)

Chromosomes do not basically differ from chromatin in their composition. According to the latest observations, chromatin has a subunit structure, its subunits, called nucleosomes, having a diameter of 7–12 nm. Nucleosomes are composed of a histone core which is wrapped up in DNA of superhelical structure. These spherical shapes are interconnected by filaments, apparently made of DNA, the resulting impression being of a string of beads. This long chain of nucleosomes is then wound into a higher helical structure which gives rise to a thick chromatin filament. The spatial orientation of chromatin and its changes are governed by enzymic phosphorylations* and acetylation of histones.

---

* The activity of some animal protein kinases is stimulated by cAMP and cGMP. Some of these enzymes are stimulated specifically by cGMP. An increased level of protein kinases is apparently associated with initiation of cell division.

A chromatin filament is coiled both longitudinally and sideways (it winds around itself), which generates the optically visualized chromatid. Every chromosome at the beginning of nuclear division is composed of two identical chromatids, the so-called sister chromatids, connected by the centromere.

The complexity of the chromosome structure with all its DNA strands is shown schematically in Fig. 26 (the number of DNA filaments is 32 before and 64 after replication). The strands are not oriented parallel to one another in the chromosome.

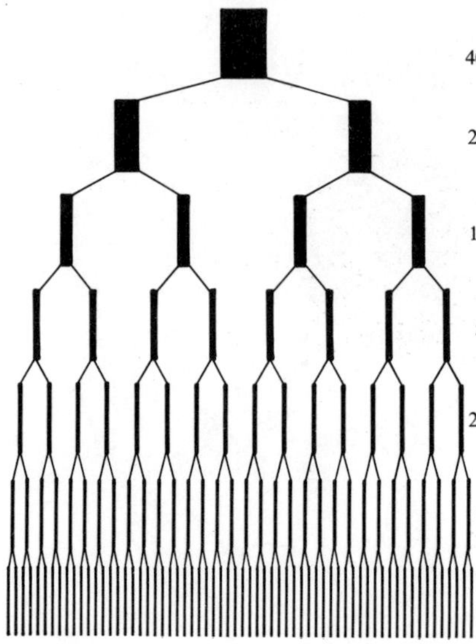

Fig. 26. Schematic representation of the fibrillar complexity of a chromosome

The shape of chromosomes is depicted in Fig. 27, showing the set of chromosomes from late anaphase of the second meiotic division in salamander sperm cells.

The number of chromosomes in diploid cells of higher organisms ranges from 2 to 500. Thus, *Drosophila melanogaster* contains 8, man 46 and crayfish 116 chromosomes in every somatic cell. The number and morphology of the chromosomes form the so-called cell karyotype. In a diploid somatic cell chromosomes occur in pairs, which has a functional reason. The paired chromosomes are of identical shape and are called homologous; a diploid cell thus contains two identical sets of chromosomes (2n). Nonhomologous chromosomes differ in size and in the position of the centromere, which appears as a constricted part on the chromosome. The functional identity or homology of paired chromosomes is based on the fact that each of the chromosomes in the pair contains an analogous allele at its appropriate site (even if it may be qualitatively different). Sexual cells formed from the diploid cells during meisosis contain only a single set of chromosomes and are called haploid (n).

Nuclei contain nucleoli. At least one nucleolus is present for every haploid set of chromosomes, there being a link from it to a specific chromosome. According to present views the nucleolus is instrumental in the synthesis of RNA, especially rRNA. Nucleoli contain DNA, RNA, basic and acid proteins, phospholipids and enzymes (RNA polymerases, RNA methylases and ribonucleases).

Fig. 27. A set of chromosomes

Nucleolar genes were observed in the electron microscope and may be seen in Fig. 28, showing genes isolated from the nucleolus of a salamander egg. These genes are involved in the process of synthesis of a great many ribosomal RNA molecules formed simultaneously or in a close sequence. The individual genes are separated by segments which are not transcribed. The longitudinal filaments (genes) are split by deoxyribonuclease while the side-chains (rRNA precursors) can be removed by ribonuclease, trypsin or pepsin. The preparation for electron microscopy was obtained by dispersing the salamander egg nucleus in deionized water. The nonextended forms were removed by centrifugation and the preparation was contrasted with phosphotungstic acid.

RNA is also present in other intracellular structures, the ribosomes. A detailed descripiton of their function in proteosynthetic processes and the characterization of ribosomal RNA are shown on p. 60. In eukaryotic cells some of the ribosomes are attached to a system of internal membranes called the endoplasmic reticulum. If the endoplasmic reticulum contains no ribosomes on its surface it is called smooth reticulum, if it carries ribosomes it is called rough reticulum.

Another important cell organelle is the Golgi apparatus, which is usually connected with the endoplasmic reticulum. The Golgi apparatus is involved in the transport of large molecules and is the site of synthesis of polysaccharides and lipoproteins.

Animal cells contain two bodies called the centrioles, which are believed to steer the movement of chromosomes during nuclear division.

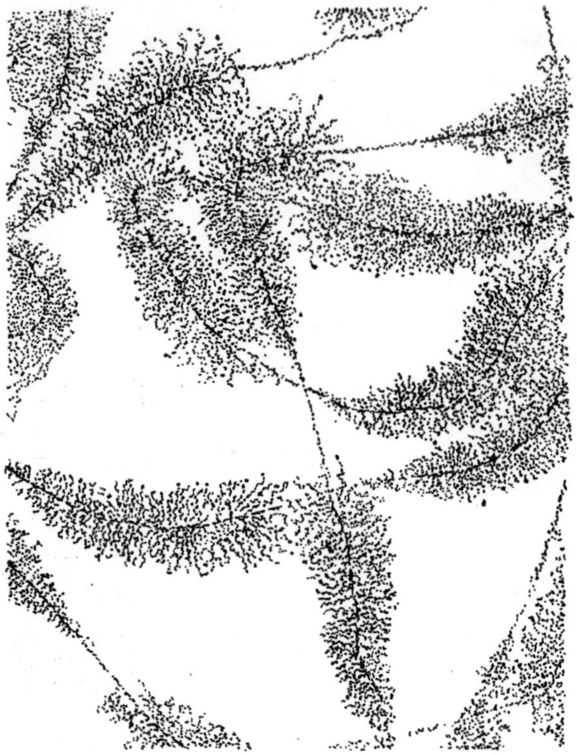

Fig. 28. Electron micrograph of rRNA synthesis on nucleolar genes

A unique organelle of eukaryotic cells is the mitochondrion. Mitochondria contain a complete protein-synthesizing apparatus (including 70S ribosomes, rRNA and regulatory factors), enzymes of the tricarboxylic acid cycle, of fatty acid oxidation, respiratory chain and oxidative phosphorylation, as well as enzymes catalysing replication and transcription of mitochondrial DNA. They are about 2 μm long and 0.5 μm in diameter and they contain about 75% protein and 25% lipid by weight, the lipids being mainly phospholipids. Mitochondrial DNA is cyclic, double-helical and does not form hybrids with nuclear DNA.* Its total content is about 1% of the corresponding nuclear DNA.

---

* It was established in rat mitochondria that mitochondrial DNA exhibits strictly matroclinic (maternal) inheritance.

Mitochondria are replicated semiautonomously, DNA synthesis being catalysed by mitochondrial polymerase. The mitochondrial genome is not sufficiently large to code for all mitochondrial proteins so that most mitochondrial proteins are in fact synthesized in the cytoplasm and are coded by nuclear genes. Only 20% of the proteins of the inner mitochondrial membrane are coded by mitochondrial DNA.

The genetic code of the mitochondrial and chloroplast genomes was found to differ slightly from that of the nuclear genome. The termination codon UGA is read in mitochondria as tryptophan and the AUA codon as methionine. The differences observed are due to the relatively small number of tRNAs coded by mitochondrial DNA (22) in comparison with the number of tRNAs coded by nuclear DNA (more than 40).

The mitochondrial DNA of *Saccharomyces cerevisiae* is roughly five times longer than mammalian mitochondrial DNA. While yeast mitochondrial DNA contains a number of repetitive sequences, such sequences have not been found in mammalian mitochondrial DNA. The genetic map of yeast mitochondrial DNA is relatively well described.

The thoroughly characterized proteins of *Saccharomyces* mitochondria include ATP synthetase and cytochrome *c* oxidase. ATP synthetase, an important enzyme of oxidative phosphorylation, is composed of some ten protein subunits, four being synthesized in mitochondria, the rest being synthesized in the cytoplasm. Cytochrome oxidase contains seven protein subunits, three being synthesized in mitochondria, four in the cytoplasm. The subunits formed in mitochondria are larger and more hydrophobic. The main role of the enzyme is the reduction of oxygen to water.

The cytoplasm contains the so-called microbodies of different types, such as lysosomes containing various hydrolytic enzymes (e.g., ribonuclease and deoxyribonuclease) and peroxisomes (including the so-called glyoxysomes), containing some oxidases and catalase.

The so-called microsomes are created as artifacts during cell fractionation and mostly represent fragments of endoplasmic reticulum.

Plant and algal cells are distinguished by the presence of chloroplasts, containing chlorophyll, carotenoids and enzymes required for the primary reactions of photosynthesis. Chloroplasts also contain circular DNA.

Molecular Mechanism of Mutations

Mutation is a sudden and permanent change of the genome. Individuals with a mutated genetic information content are designated as mutants.

The mutation rate reflects the probability with which a certain mutation event takes place in the examined biological unit (virus, cell, organism) in one generation. The mutation rate is usually defined from the mutation frequency.

In classical genetics mutation was characterized by a hereditary change of the phenotype. Now that the material basis of hereditary properties is understood, mutation may be defined as any sudden, genetically based change of biologically active DNA (RNA in some viruses and phages).

Mutations take place in nature spontaneously and doubtless represent one of the important factors in the evolution of life. However, mutations can also be brought about by mutagens, either physical or chemical.

Action of mutagenic factors may occasionally result in profound alterations of the genetic structure so that the affected cell cannot fulfil its basic-life functions and eventually dies. Such a mutation is called lethal.

Changes in the number and shape of chromosomes are called chromosomal mutations. Changes affecting a single gene are gene mutations. It follows from the foregoing text that a certain segment of DNA (gene) may control the synthesis of a certain enzyme. If such a DNA segment or its part is lost, the cell is unable to synthesize the enzyme. In fact, the enzyme may not be synthesized even after a single nucleotide pair has been altered. This is then a point mutation.

A point mutation can give rise to a new code triplet which happens to code for the same amino acid and then an identical enzyme is synthesized, or to a triplet for another amino acid which may result in an altered biological activity of the protein product, or to a nonsense (terminating) triplet when an incomplete polypeptide chain is formed.

Another type of mutation is the frameshift mutations caused by either adding or deleting a nucleotide pair in the DNA strand. Owing to this change, all subsequent triplets are read differently.

Mutation which eliminates the function of a certain gene is called a forward mutation. Recovery of the original function due to another mutation is called reversion. Reversion thus results in the restoration of the original phenotype.

Chemical Mutagenesis

Use of chemical agents for bringing about mutational changes has replaced to a great extent the earlier physical mutagens (e.g., UV light). The application of physical mutagens often resulted in cell death, i.e. a lethal mutation. Chemical mutagens can increase the mutation rate as much as 10^6 times without affecting cell viability.

In point mutations the base pairs are altered in two ways:

a) if the base pair in the DNA double helix retains the orientation of purine-pyrimidine or pyrimidine-purine, this represents a simple replacement or transition as shown in scheme [46];

[46]
AT ⟶ GC
GC ⟶ AT
TA ⟶ CG
CG ⟶ TA

b) if the orientation of purine-pyrimidine or pyrimidine-purine is changed we are dealing with a complex replacement or transversion, as shown in scheme [47].

[47]
AT ⟶ TA
⟶ CG
GC ⟶ CG
⟶ TA

Let us examine the possibilities of a transition caused by the use of base analogues. 5-Bromouracil is mostly incorporated into DNA in the place of thymine and during reduplication of DNA pairs with adenine. Occasionally, however, pairing with guanine may take place. This incorrect pairing can be caused by a change of the tautomeric state of the pyrimidine base as the bromine atom is introduced, bromine being more electronegative than the methyl group. The consequence of such an error is shown in scheme [48], where every pair of bases gives rise to two daughter pairs during replication.

[48]

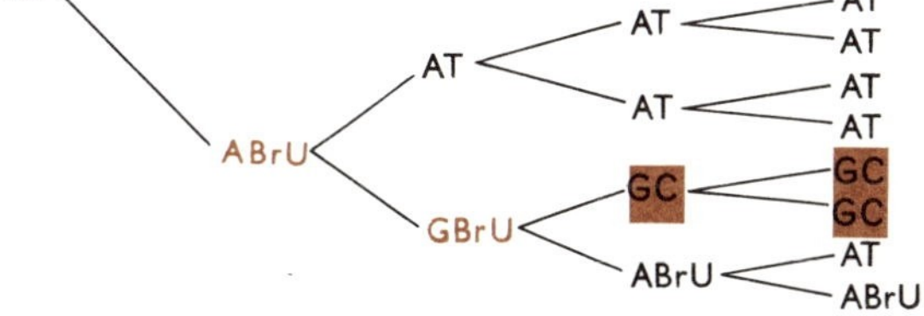

Scheme then shows how the formation of the incorrect pair of bromouracil with guanine causes transition.

More recently, the mutagenic effect of 5-bromouracil has been explained in a different way. Incorporated 5-bromouracil is first dehalogenated to uracil. This is removed from DNA through the action of a specific glycosylase and the apyrimidine site thus formed is then excised enzymically. Subsequent repair synthesis restores the double-helical nature of the strands and it is under these conditions that the repair system can make an error which results in transition [49].

2-Aminopurine is usually incorporated into DNA in the place of adenine as it forms two hydrogen bonds with thymine just as adenine does. In contrast with adenine, 2-aminopurine can also form a hydrogen bond with cytosine which again results in transition – see scheme [49].

[49] AT $\longrightarrow$ GC

Transversions can arise through purine-purine or pyrimidine-pyrimidine pairing in the presence of substances which permit such anomalous pairing by interfering with DNA structure, or which cause a deletion of base pairs followed by addition of incorrect bases. A typical compound causing transversions is proflavin [50].

[50]

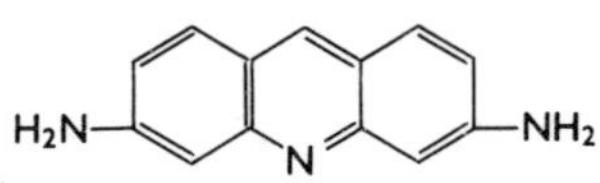

Mutants generated with the aid of 5-bromouracil can be made to revert by repeated application of the mutagen. If a given bacterial strain lost the ability to synthesize histidine to the mutagenic action of 5-bromouracil, the resulting inability is now a phenotypic manifestation of the mutational change of the strain genotype. Exposure of such nutritional mutants to 5-bromouracil can restore the ability to synthesize histidine by a part of the cells.

The cause of the reversion can consist in changing the base pair damaged during the first (point) mutation*. Even if this were the simplest explanation for the phenomenon it is apparently not the most likely one. Gene function can be restored even by other mechanisms of reversion. Thus, e.g., the first mutation results in a nonsense triplet that does not code for any amino acid, or in a triplet coding for an incorrect amino acid. Reversion is then brought about by a replacement of a base in other pairs of the same triplet in which the first mutation has taken place. Owing to the second mutation a code triplet can be formed that does not code for the original amino acid but the enzyme formed can at least possess partial enzyme activity.

---

* Reversions can take place not only through altering a base pair but also through insertion or deletion of a base pair.

The cause of reversion may also be sought in the so-called suppressor mutation when one mutation is eliminated by another, caused at a different locus of the chromosome. Suppressor mutation can be intragenic (both the first and the second mutations are localized on the same locus but not in the same code triplet) or intergenic (the two mutations occur on different genes). The gene which suppresses a mutation in another gene is called a suppressor gene.

Intragenic suppressor mutations are often caused by insertion or deletion of a nucleotide pair. Occasionally, intragenic suppression is caused by a mutation resulting in the replacement of an amino acid in the polypetide chain. Incorporation of the "new" amino acid can bring the conformation of the protein (enzyme) damaged by the first mutation at a completely different site, back to the functional form.

In intergenic suppressor mutations the structure of the mutated gene does not change but the manner of mRNA translation is affected. Thus, a mutation in the tryptophan synthetase gene in *Escherichia coli* resulted in the replacement of glycine with arginine. A molecule with this amino acid at that particular position has no enzymic activity. Suppressor mutation resulted in a new mutant where the mutated codon was again read as glycine. It could be shown that the suppressor mutation affected the segment of DNA coding for arginine tRNA in such a way that the "new" suppressor tRNA incorporated glycine into the polypeptide molecule. In *Escherichia coli* there are three different genes coding for $tRNA^{Arg}$; only one of them was mutated so that the other two could control the synthesis of $tRNA^{Arg}$.

Suppressor genes were found to exist both for codons coding for the incorrect amino acid and for stop codons.

Suppressor genes can result in false reading of normal genes and produce "wrong" proteins, but it appears that such a loss is not crucial for the cell as sufficient correct copies of the protein are still formed.

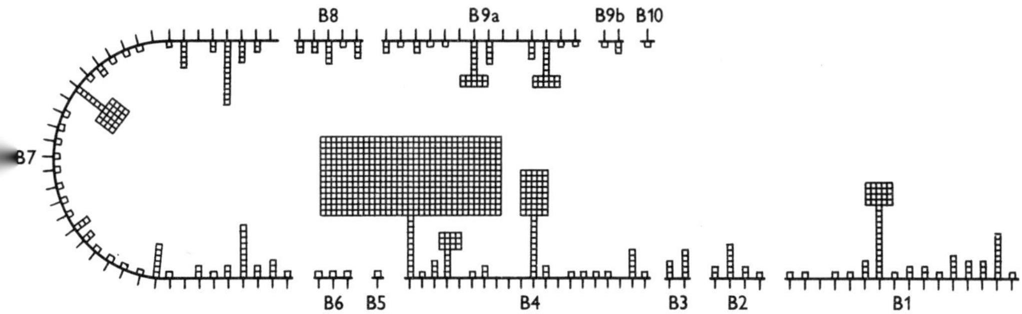

Fig. 29. Hot spots in the rII region of bacteriophage T4

Before proceeding to the explanation of the mechanism of action of the individual important mutagens, something should be said about the fact that point mutations are not randomly distributed over the entire gene. Using the example of the rII region of bacteriophage T4, Benzer showed that the gene contains sites at which point mutations are much more frequent than elsewhere and called them hot spots. Fig. 29 shows the mutation spectrum of the rII region (the B cistron). Every square represents an independently formed mutation. It will be seen that there is one particular site in the region which is "prone" to mutation and many sites where no mutation has ever been observed.

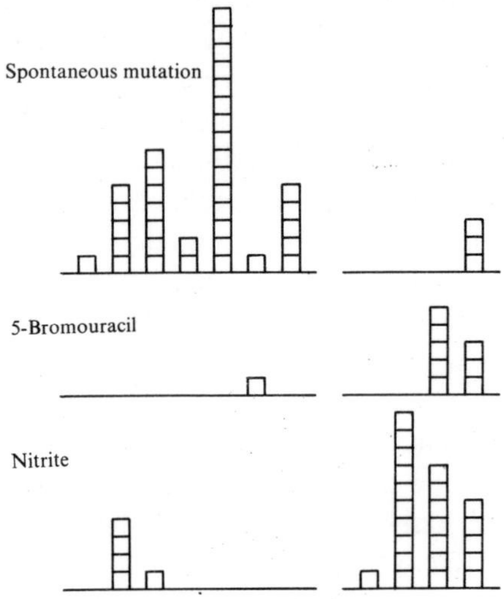

Fig. 30. Distribution of hot spots depending on the type of mutagenesis

The hot spots have a different distribution, depending on the type of mutagen. Again using data obtained by Benzer, two short segments of the rIIA gene are shown in Fig. 30. Every type of mutagen prefers different sites in the DNA molecule.

The exceptional sensitivity of the hot spots to both spontaneous and induced mutation is most probably due to the nature of bases which lie adjacent to the hot spots in DNA.

5-Bromouracil and 2-Aminopurine

The mutagenic effect of 5-bromouracil probably consists in random incorrect pairing of this base with guanine as 5-bromouracil is correctly incorporated in DNA at the place of thymine and would thus pair with adenine. The consequences of this error were discussed previously. It was shown that mutagenesis by 5-bromouracil really requires the incorporation of this antimetabolite into DNA; even at a high degree of incorporation of 5-bromouracil into DNA the nearest neighbour technique (p. 48) was unable to demonstrate a change in the base sequence. According to the latest observations, the mutagenic effect of 5-bromouracil is caused by a complex molecular machinery, including dehalogenation of incorporated 5-bromouracil, excision of the uracil residue from DNA and subsequent repair. Model experiments with a copolymer of deoxyadenylic acid and deoxy-5-bromouridylic acid showed that during replication under catalysis of DNA polymerase deoxyguanylic acid can be incorporated into the nascent DNA. Such a phenomenon has never been observed during analogous experiments with a copolymer of deoxyadenylic acid and thymidylic acid.

The purine analogue 2-aminopurine also results in mutations of the transition type. Here again the antimetabolite is incorporated into DNA. Even without resorting to a tautomeric shift in its molecule, the structure of 2-aminopurine permits the formation of two hydrogen bonds with thymine and one hydrogen bond with cytosine; this second possibility results in the transition.

Alkylating Agents

This group includes dimethyl sulphate, diethyl sulphate, ethyl methanesulphonate and ethyl ethanesulphonate [51]. These agents cause mutagenesis through two distinct mechanisms; (a) they can induce erroneous base pairing and (b) they result in errors during the repair processes. In DNA, the alkylating agents will alkylate N1, N^2, N3, N7, C8 and O^6 in guanine, N1, N3, N^6, N7 and C8 in adenine, N3 and N^4 in cytosine and N3 and O^4 in thymine*. Phosphate triesters are also formed. The 7-alkylguanine residues, which represent the main alkylation products, probably do not play a major role in the mechanism of mutagenesis. O^6-Alkylguanine and O^4-alkylthymine may cause errors in pairing; the other alkylations perhaps result in a loss of purine or pyrimidine residues or induce repair processes. However, it was shown recently that depurination and depyrimidination have lethal rather than mutagenic consequences.

[51]

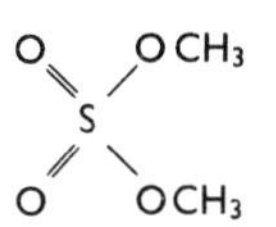

Dimethyl sulphate

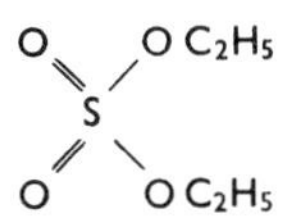

Diethyl sulphate

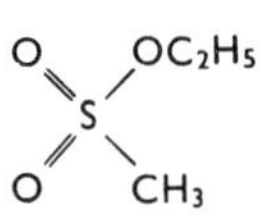

Ethyl methanesulphonate (EMS)

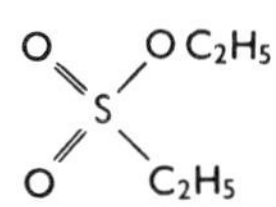

Ethylethane sulphonate

* The number next to the chemical symbol indicates substitution directly on the atom; the number raised as a superscript indicates substitution in a group attached to the atom.

Another important alkylating agent is N-methyl-N′-nitro-N-nitrosoguanidine:

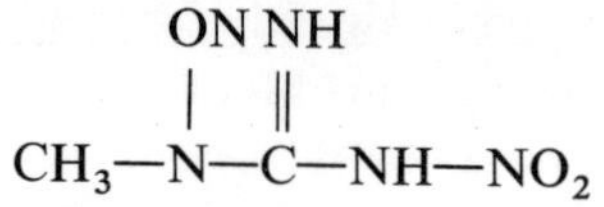

This brings about mutations in bacteria, yeasts, algae, plants and fruit-flies. It has antimicrobial and cancerostatic properties but occasionally it may act as a carcinogen. It will alkylate DNA, giving rise to 7-methylguanine as the principal product. It also reacts with nucleophilic groups, such as the ε-group of the lysine residues of proteins. N-Methyl-N′-nitro-N-nitrosoguanidine prefers in its attack replicating DNA, where it causes multiple closely linked mutations in the replication fork.

Nitrous Acid

Nitrous acid causes deamination of guanine, cytosine, hydroxymethylcytosine and adenine residues in nucleic acids. As a consequence of these deaminations, transitions will appear, a change of the GC pair to the AT pair being more common than a change of AT to GC. Deamination of guanine leads to xanthine, which retains its ability to pair with cytosine. On the other hand, cytosine gives rise to uracil, which can be removed from DNA by a specific glycosylase. Nitrous acid can also create cross-links which prevent separation of DNA strands and thus block replication. This nitrous acid effect probably causes large deletions, as were observed in the rII region of the T4 bacteriophage. Nitrous acid was used for inducing mutants of the tobacco mosaic virus.

Hydroxylamine

The action of hydroxylamine on RNA at higher pH values causes destruction of uracil. At lower pH values hydroxylamine reacts with a cytosine residue. Thymine in DNA reacts with this mutagen only to a small extent. The reaction of hydroxylamine with cytosine in nucleic acids is shown by equation [52].

[52]

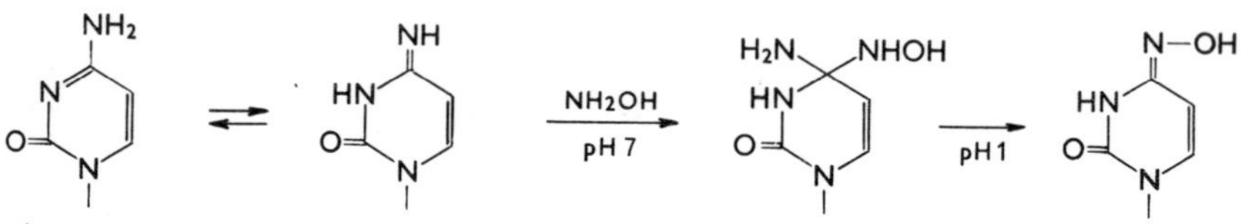

As the NH-group appears in position 3 of the pyrimidine ring, the possibility of pairing with adenine exists and a transition from GC to AT ensues.

Acridines

Acridine dyes cause frameshift mutations during DNA replication (deletion or addition of individual base pairs). In the presence of acridines the double helix of DNA is extended and the acridine molecule is intercalated between two layers of base pairs.

Intercalation deforms the DNA double helix and creates conditions for addition and deletion. The action of acridine compounds gives rise to mutations with a complex interchange of bases (transversion) which do not revert under the influence of base analogues. Intercalation can occur both in DNA and RNA double helices.

The above enumerated compounds do not cover the whole field of mutagens. Other mutagens have been used, such as mustard gas, urethane, nitrosamines, triethylenemelamin, cyclophosphamide, N-methyl-N-nitrosourea, formaldehyde, peroxides, ketene, N-bromosuccinimide, aflatoxins, mitomycin C and many others. The mechanism of their action may be polyvalent but it can be fitted into one of the above categories.

Directed Mutagenesis Using DNA Recombination Techniques

Modern methods based on sequencing and preparation of recombinant DNA molecules provide us with new possibilities in the preparation of mutants. For instance, a selected DNA segment can be deleted if it is located between two sites cleaved by two restrictases (see p. 157). Another possibility lies in preparing a substitution mutant with the aid of a synthetic oligodeoxyribonucleotide of known base sequence. This oligodeoxyribonucleotide is complementary with a certain DNA sequence but it contains one or two intentionally altered bases. If it is mixed with a cloned complementary strand it will form a hybrid (the few replaced bases will not affect the hybridization). Under catalysis of DNA polymerase I the attached oligodeoxyribonucleotide is made up to a complete strand. This "artificial" gene is then introduced into a bacterial cell where it replicates, giving rise to two different molecules of DNA, the original and the mutated one.

Radiation and Mutation

Ionizing Radiation

If ionizing radiation (α, β and γ radiation, X-rays, neutrons) is applied to microbial populations, the number of surviving cells depends on the radiation dose. Curves expressing the dependence between microbial survival and radiation dose are exponential in character (Fig. 31).

The number of mutants in a surviving population is usually small and is directly proportional to the radiation dose.

According to the original concept of the mutagenic effect of ionizing radiation the mutation was assumed to ensue from primary ionization of the sensitive site in

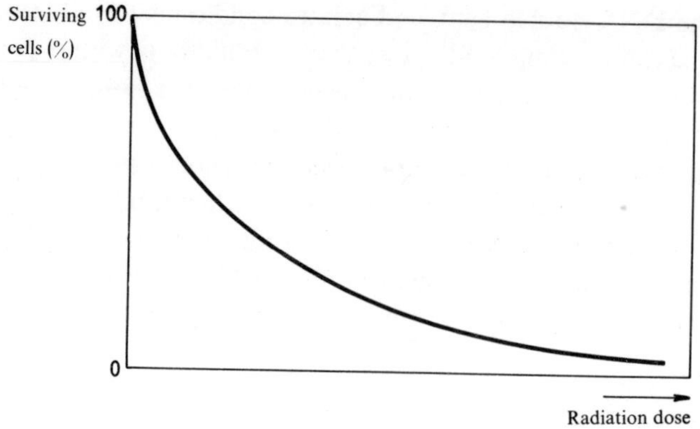

Fig. 31. Dependence of cell survival in a microbial population on the dose of ionizing radiation

genetic material. This is called the target theory. According to present views, ionizing radiation first causes the formation of free radicals and excited states in both water and DNA. The concentration of superoxide radicals formed in living cells (they are also formed by UV radiation) is considerably decreased by the recently discovered metalloenzyme superoxide dismutase, which is present in all respiring cells.

Following the action of X-rays or of hydrogen peroxide, bacterial DNA was found to contain adenine N-oxide. Conversion of adenine to the oxide strongly affects pairing with thymine.

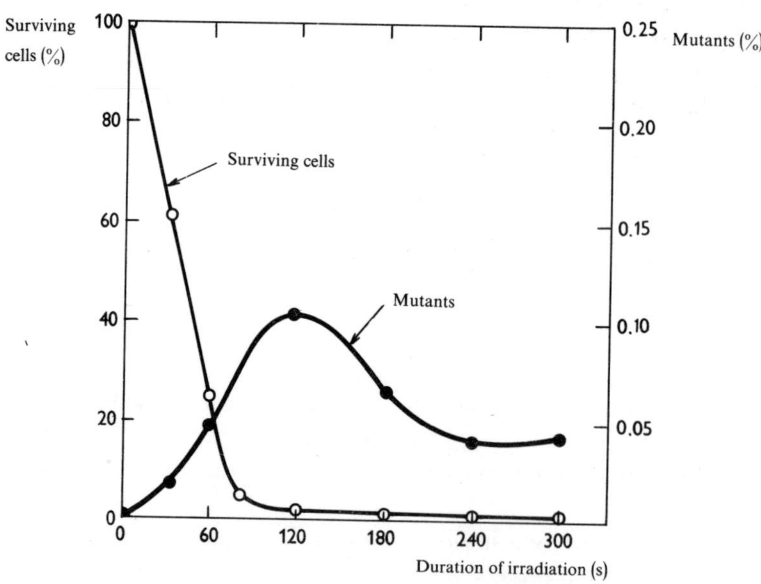

Fig. 32. Dependence of cell survival in a microbial population and of mutation frequency on the length of exposure to UV light

UV Radiation

While the dependence between the ionizing radiation dose and mutagenic effects may be roughly linear in a certain dose range, with UV light the mutagenic effect shows a maximum at a certain dose (Fig. 32).

The effect of UV light depends on many physiological, chemical and physico-chemical factors.

When thymine solutions in water are illuminated with UV light, the absorbance of the solution at 260 nm decreases and a new substance appears in amounts corresponding to the absorbance decrease. If frozen solutions of thymine are exposed to UV light the photoproduct is formed in a particularly high yield. A substance similar to this photo-product could also be isolated from DNA of bacteria irradiated with UV light. Analysis of the substance showed it to be a thymine dimer [53] which is relatively resistant to the action of acids and hydroxides.

[53]

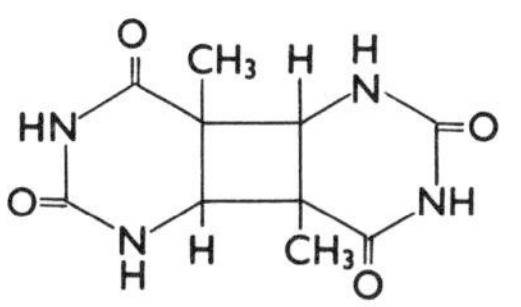

During irradiation of bacteria some 10–16% of thymine residues in DNA are dimerized; the structure of a thymine dimer is shown in [54]. As vicinal thymines are most able to dimerize, the process can take place only with a portion of the thymine complement of DNA where just 10–16% of the sequence is pTpTp. UV illumination also affects residues of uracil in RNA where uracil residues may dimerize or molecules of water may be added to the double bond of the pyrimidine ring in position 5–6.

[54]

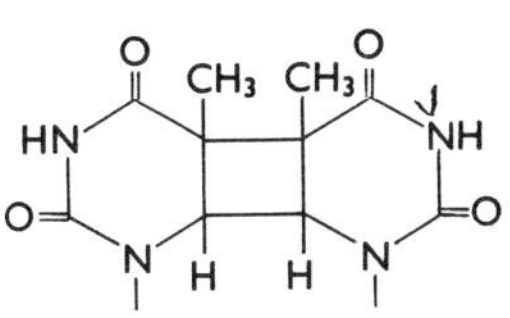

Cytosine can also undergo dimerization, the resulting dimer readily undergoing deamination with formation of uracil dimers. However, the main photo-product here is the adduct with water. The number of cells surviving UV illumination is indirectly proportional to the content of thymine dimers in DNA. The mechanisms permitting the cell to repair the effects of UV light on DNA were described above (p. 45). Fig. 33 summarizes schematically the effects of UV light on DNA structure.

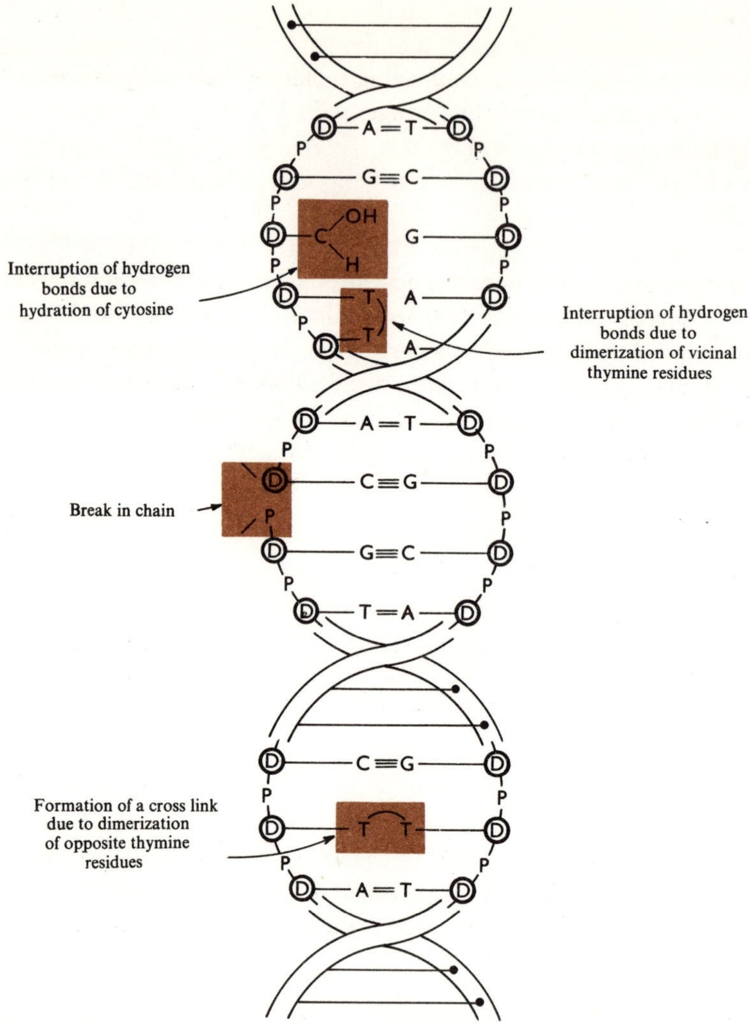

Fig. 33. Schematic representation of the effects of irradiating DNA with UV light. D, Deoxyribose; P, phosphoric acid

It was found that mutation frequency is significantly affected by the ratio of 3′→5′ to 5′→3′ nuclease activity of the corresponding DNA polymerase, i.e., the ratio of "repair" activity to the "polymerization" activity. If the 3′→5′ nuclease activity is low the mutation frequency is low. This fact is connected with the important role of 3′→5′ exonuclease activity of DNA polymerases in the repair mechanisms compensating for "wrong" bases.

Mutations represent an exceptionally important factor of evolution. A high mutation frequency is not desirable from the evolutionary point of view as it often results in progeny that are not viable. In the other extreme, when too few mutations occur there is not a sufficient selection of mutants to cope with and adapt to changing environmental conditions.

Integrated Segments, Transposons and Jumping Genes

It was found only recently that the galactose and lactose operons of bacteria can undergo special mutations which cannot be explained by previously known principles. The reversion frequency in these mutation types is extremely low and is not affected by mutagenic factors. It was found that these mutations are caused by insertion of new DNA segments into the existing operon (some 1 000 nucleotide pairs being introduced). The inserted segments were called integrated segments and abbreviated IS. The presence of integrated segments can influence the function of neighbouring genes or cause deletions. Such integrated segments were found in bacteria, some plasmids and phages.

It will be seen later (p. 136) that resistance to antibiotics and other inhibitors can be caused by the so-called R plasmids. Some genes of these R plasmids are incorporated in the bacterial chromosome and it can happen that the incorporation takes place within a gene which is obviously inactivated by it. The incorporated genes of R plasmids contain integrated segments at both ends (so-called insertion sequences); these incorporated sections are called transposons.

Insertion sequences also bring about a great mobility of the genes at whose ends they occur; such "jumping genes" can then rapidly spread among bacterial populations.

Application of Mutagens and Selection of Mutants

General Aspects

The starting biological material for the preparation of microbial mutants is called the parent strain. If such parent strains were originally isolated from nature they are called wild strains and, likewise, their genotype is the wild genotype. Microorganisms are an extraordinarily suitable object from which to derive various mutants, primarily because bacteria are haploid and so are a great many other microorganisms at one or other stage of their life cycle. In the haploid state a mutational change is relatively easy to detect as there is no pair gene (allele) that would at least partly compensate for the lost gene in a diploid cell.

If a mutation takes place in a haploid cell, the cell grows and divides to give rise to a mutant clone. If the cell is multinuclear, the mutation produces a heterokaryon in only one of the haploid nuclei (p. 183). Such a cell grows on a solid medium into a sector colony where only one sector of the circular growth arises from the mutant clone. It is for such reasons that mutants are generally prepared during the haploid stage of the parent microorganism and, whenever possible, during the stage when the cell contains a single nucleus (in the case of bacteria and antinomycetes a single nuclear equivalent). This condition is fulfilled by fungal or actinomycete spores and by vegetative as well as spore stages of bacteria.

Microorganisms may undergo spontaneous mutations, albeit at a very low frequency ($10^{-9}-10^{-7}$). Application of a mutagen may increase the number of mutants in a surviving microbial population by several orders of magnitude; in such cases we speak of an induced mutation. At a high intensity of application, mutagens will kill most of the cells, the damage being so extensive that the cells cannot perform basic metabolic functions and hence die. It is usual to distinguish between the so-called "genetic killing", which includes lethal mutations due to damage of the genetic equipment of the cell, and "nongenetic killing", which is due to the general toxicity of the mutagen. With haploid cells a mutation may become lethal if it causes nutritional dependence on a particular substance and the mutant is then cultivated in a minimal growth medium. The mutant is unable to grow and it dies (a relatively lethal mutation). In such cases the classification of mutants into lethal and viable depends on the growth conditions subsequently employed. For practical application it is preferable to work with mutagens possessing low nongenetic killing.

With haploid unicellular organisms one can readily determine the dependence of the frequency of viable mutants on the mutagen dose, on the one hand, and of the fraction of surviving cells on the mutagen dose, on the other. It is true for all mutagens that the fraction of viable mutants in a surviving population increases with increasing dose only up to a certain level.

For every mutagen and every organism it is necessary to check experimentally the optimal mutagen dose at which the fraction of mutants in the surviving population is highest. The optimal concentration of a mutagen obviously depends on the duration of its application and on various other conditions. The optimal conditions may differ even for the same mutagen if different categories of mutants are sought. If potent mutagens are employed it is usually not expedient to work in the range of highest mutant number; drastic changes in DNA may occur and hence a high percentage of mutants with a complicated genotype may arise.

The choice of a mutagen usually proceeds from empirical tests. For instance, if a wide spectrum of auxotrophic mutants is to be obtained by a single operation, a more potent chemical mutagen is preferable. If mutants are desired with a deficiency of a particular enzyme, it is advisable to employ two or more mutagens in sequence at lower doses. To increase the production of secondary metabolites it is expedient to apply low doses of less potent mutagens (such as UV light or nitrous acid) in combination with a repeated selection of mutants. It is then necessary to check the variability of production of secondary metabolites within the populations obtained.

Mutants may differ from parent strains in a number of phenotypic features; for example, in their nutritional dependence (auxotrophic mutants), changes in regulatory mechanisms (e.g., constitutive mutants), acquisition of resistance to inhibitors (e.g., to antibiotics) and phages, in temperature sensitivity, in the pigmentation of colonies or in the morphology of cells or colonies. For biotechnological applications mutants capable of producing high yields of amino acids and of secondary metabolites, or mutants containing a high level of a desired enzyme, are frequently sought.

Technical Aspects of Work with Mutagens

Induction of Mutants by Irradiation

One of the earliest and still employed mutagens is ultraviolet light. The source used is generally a low-pressure discharge tube (e.g., a 15 W Hanovia germicidal lamp) emitting 90% of its energy in the range of maximum absorbance by nucleic acids (254 nm). Sources of UV light are calibrated by measuring their intensity in $J\ m^{-2}s^{-1}$, or in a biological test based, for instance, on inactivation of bacteriophage T2. The radiation is applied to a suspension of cells or spores in a physiological or buffered solution over time periods of several seconds to several minutes. It is not advisable to irradiate suspensions in rich media, such as meat-peptone broth, which protect the cells and, moreover, favour the production of toxic organic peroxides. The suspension density should be about 10^8 cells or spores per ml. The results may be affected by the physiological state of the culture.

The irradiated suspension is suitably diluted with physiological saline and is inoculated onto agar plates in Petri dishes. An important indicator of the successful procedure is the so-called survival curve, which is constructed by plotting the logarithm of the fraction of surviving cells against the UV radiation dose. The shape of the survival curve for most haploid cells is exponential. The proportion of mutants in the surviving population increases roughly linearly with the radiation dose while it slowly decreases when the proportion of survivals is low (see Fig. 32). It is common to apply a UV dose at which 0.1 – 1.0% of the population survives.

It was mentioned before that visible light induces a photoreactivation process which diminishes the mutagenic effects of UV radiation. To ensure reproducibility of the results it is recommended to treat UV-irradiated suspensions with yellow light only (sodium lamp or a yellow-coloured incandescent bulb).

Ionizing radiation will induce mutants only in special cases. It is mainly applied to microorganisms that are opaque to UV light in their haploid phase, such as is the case with heavily pigmented spores. In view of the high frequency of chromosomal breaks, application of ionizing radiation is not particularly suitable for routine work with microorganisms.

Induction of Mutants by Chemical Mutagens

A great many mutagenic compounds are now known (p. 91) but only a few have found their way into microbiological practice. The most common ones are alkylating agents, especially ethyl methanesulphonate (EMS), ethyl ethanesulphonate (EES), diethyl sulphate (DES) and N-methyl-N′-nitro-N-nitrosoguanidine (MNNG).

The first three have proved their worth as universally applicable mutagens, exhibiting a high mutagenic effect with low lethal results. They are usually applied to cell suspensions in a neutral buffer at concentrations up to 1%. The concentration of the mutagen and the length of exposure depend on the species and form of microorganism, such as bacteria, yeasts or spores. The exposure is terminated by adding sodium thiosulphate or by diluting the suspension. Successful application of these alkylating agents is closely linked to their chemical purity. If the preparations contain some by-products their lethal effect may increase while mutagenesis may be less effective.

N-Methyl-N′-nitro-N-nitrosoguanidine is distinguished by its low lethal effect and the high proportion of mutants produced. Its mutagenic effects are well understood in *Escherichia coli,* where the yields of auxotrophic mutants may attain as much as 50% of the surviving population. The compound has been variously applied at concentrations between 0.1 and 1.0 mg ml^{-1} at 37 °C for 15 – 30 min. When applied to growing cells its lethal effects will increase while its mutagenic effects will diminish. After exposure to it cells are separated from the mutagen by filtration or centrifugation.

With other microorganisms the maximum proportion of mutants rarely exceeds 10%. A case in point is a 120-min application of 3 mg of N-methyl-N′-nitro-N-nitrosoguanidine in a buffer to streptomycete spores at 30 °C. With this mutagen it is of special importance to check the optimal application conditions for every particular microbial strain. The survival curve is often examined as a function of mutagen concentration rather than of the duration of exposure to it. It is more practicable to apply this mutagen under conditions when the mutant fraction represents only a small fraction of the optimum yield as the probability of generation of undesirable multiple mutants is thus diminished.

One of the mildest mutagens is nitrous acid. It is applied as a freshly prepared aqueous solution of sodium nitrite at concentrations of $0.1 - 0.2$ mol l^{-1} in an acidic medium. Exposure to the mutagen is terminated by addition of sodium hydroxide or a neutral buffer. The survival curve is exponential after an initial lag; the fraction of mutants in the surviving population increases linearly with increasing duration of exposure.

Nitrogen base analogues and acridine dyes are not often used in practical work. As the application of base or nucleoside analogues requires special conditions, they are usually employed only in basic research. Acridine dyes have found application with bacteriophages, less so with bacteria and other microbial species. An interesting feature of acridine dyes is their ability to eliminate episomal plasmids from growing cultures (p. 145), a particularly sensitive plasmid being the F factor.

When working with chemical mutagens it is necessary to observe all the safety precautions prescribed for work with toxic substances. Alkylating agents that are liquid are never pipetted directly. In working with N-methyl-N′-nitro-N-nitrosoguanidine it is essential that the substance and its solutions are kept away from the skin and epithelia.

Selection and Isolation of Mutants

Selection techniques are based on differences between the mutant and the parent phenotypes. The most common technique is selection on solid nutrient media in Petri dishes. The composition of these nutrient media can represent the desired selection pressure. The suspension undergoing selection is inoculated on to dishes in such a concentration that individual cells would give rise to independent solitary colonies. In some cases a mutant clone can be distinguished directly by its morphology (shape and colour of colonies) or by growth in the presence of specific factors contained in the medium.

It is more complicated to select mutants that will produce different amounts of a certain compound. A direct distinction on agar plates is not possible, high-production strands being usually selected randomly. Thus, differences in antibiotic production are investigated in every isolated colony after submerged cultivation; the isolate with the highest production is then used for further improvement.

As screening of industrial strains based on random selection is time-consuming and may not yield any results, more rational ways are being sought to select and isolate high-production industrial strains. A knowledge of biochemical enzyme sequences and regulation is useful here. Another example of the selection of mutants with a high production of a certain enzyme is the use of continuous cultivation in a chemostat. Continuous selection is facilitated here by suitable modifications of the liquid nutrient medium.

Some Important Categories of Mutants

Strains capable of synthesizing all substances required for growth from a simple source of carbon (such as glucose) and from inorganic salts are designated prototrophic. A bacterial strain, the growth of which depends on the addition of a compound to the so-called synthetic minimal medium, is called auxotrophic. The cause of auxotrophy is a genetically fixed loss of the ability to synthesize an enzyme involved in the biosynthesis of a metabolite which is essential for the growth of the auxotroph.

To give an example, the prototrophic strain *Escherichia coli* K12 grows on a minimal medium containing glucose, ammonium sulphate, phosphate, citrate and magnesium. Its auxotrophic mutants which do not grow on this medium may be cultivated on a complete medium which contains casein hydrolysate as a source of amino acids and yeast extract as a rich source of a number of growth-promoting factors.

On a complete agar medium all cells that have survived the mutagenic treatment will give rise to colonies. However, if these colonies are transferred to a minimal agar medium only prototrophic strains will continue to grow. If the isolation of auxotrophic strains is to be successful every colony must originate from a single cell.

The practical selection of auxotrophic strains is usually carried out by the replica method. Colonies grown on a complete agar medium are transferred again to a complete agar medium (if the frequency of mutants is low in the surviving population the first growth of colonies is replica-plated directly). After growth of colonies their replica is transferred to other agar plates using a replicating stencil covered with velvetten*. A minimal as well as a complete agar medium is thus replica-plated. The principle of the method can be seen in Fig. 34.

It is common practice to determine the so-called group dependence of auxotrophic mutants. Replica-plating can be used again in such a way that replicas are transferred to minimal medium containing (a) casein hydrolysate without vitamins, (b) nucleic acid hydrolysate with addition of thymine and (c) water-soluble vitamins.

The result of the test will show whether the growth of the auxotrophic mutant depends on the presence of an amino acid, a precursor of nucleic acid or a water-soluble vitamin. In a later phase of plating it is possible to determine whether some individual substance or a group of substances in the mixtures used underlies the nutritional dependence of the strain. Crystals of amino acids, nucleic acid precursors or vitamins are placed on the surface of a minimal agar medium containing the auxotrophic strain;

---

* An all-metal replicator was developed in this laboratory, with a number of advantages over the classical velveteen type especially for serial work (Folia Microbiol. **23**, 27, 1978).

growth in the immediate vicinity of one of the crystals will indicate the dependence of the auxotrophic strain on that particular substance.

In some cases even after a mutagenic treatment the affected gene can retain a certain information activity. The genetic block caused by such a leaky mutation is incomplete and the affected gene can code a protein with a certain residual enzymic activity (a "leaky" protein).

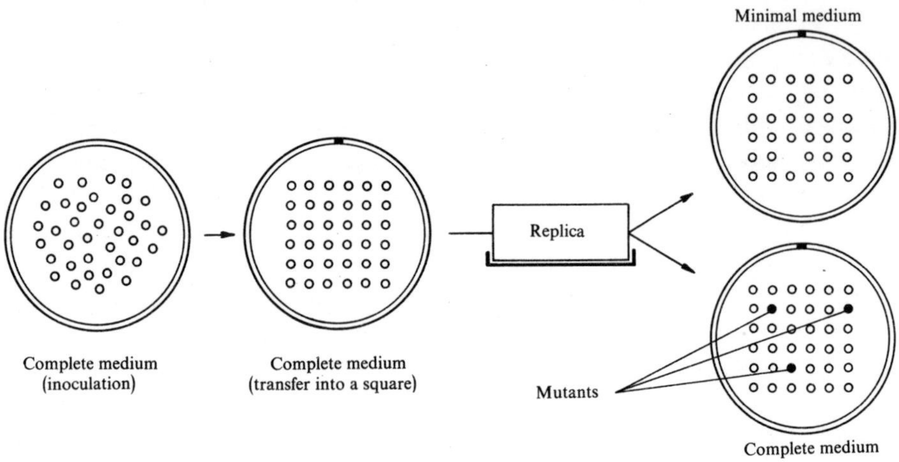

Fig. 34. Technique of isolating auxotrophic mutants of microorganisms

If the incidence of auxotrophic mutants in the surviving microbial population is too low, some enrichment method is in order. The most common enrichment technique is the penicillin method, which is based on the fact that penicillin will only affect growing cells where it inhibits cell wall synthesis. On transferring cells exposed to a mutagen to a liquid minimal medium containing penicillin and incubating them for some time in a thermostat only prototrophic cells will grow in the minimal medium and these will perish through the action of the antibiotic. This results in a substantial increase in the fraction of auxotrophic mutants. The method has the serious drawback that the correct dose of the antibiotic must be determined for every strain. The principle of the penicillin method is depicted in Fig. 35.

Auxotrophic mutants have played an important role in biochemical genetics, as may be seen from the following example.

A number of mutants have been isolated and found to lack the ability to synthesize a certain metabolite D. Detailed analysis revealed that the metabolite could be replaced in some of the mutants with substance C; mutants of this group could grow in the presence of either D or C. It goes without saying that C and D were probably related compounds, members of the same metabolic pathway. Another group of mutants was found also to grow in the presence of B or of A. The mutants could thus be divided into the following groups:

1. strains exclusively requiring substance D for growth;
2. strains requiring either D or C for growth;
3. strains requiring either D or C or B for growth;
4. strains requiring either D or C or B or A for growth.

Correct interpretation of this fact leads to a metabolic scheme where metabolite D is formed by the following sequence of reactions:

A → B → C → D

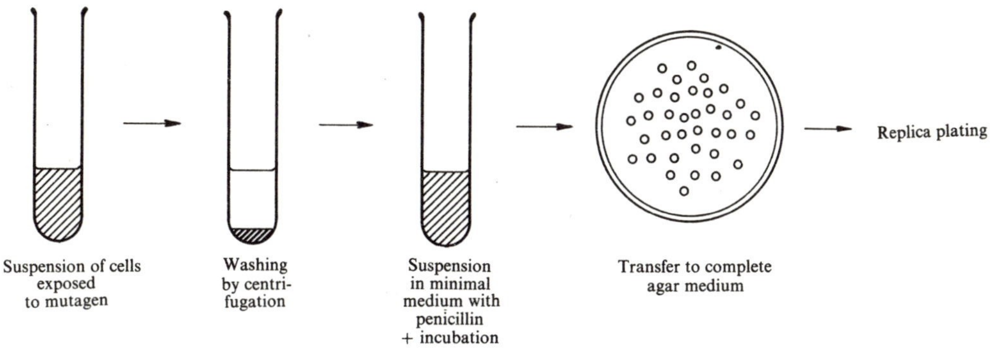

Fig. 35. Principle of the penicillin enrichment method

It was on this principle that the biosynthesis of many essential metabolites was elucidated, first of all amino acids. Clarification of the mechanism of tryptophan synthesis from anthranilic acid in *Salmonella typhimurium* was aided by the isolation of a number of auxotrophic mutants requiring tryptophan for growth. The analysis followed exactly the above pattern: in group A mutants tryptophan could be replaced with anthranilic acid, 3-indolylglycerophosphate or indole; in group B tryptophan could be replaced with 3-indolylglycerophosphate or indole; in group C the deficiency could only be compensated by tryptophan or indole; group D required strictly tryptophan. In group A cells none of the precursors would accumulate, in group B the anthranilic acid level increased, in group C the 3-indolylglycerophosphate level increased while in group D both the indole and 3-indolylglycerophosphate levels increased. It was further shown that nutritional complementation exists between these auxotrophic mutants during mixed cultivation. If it is assumed that in the Trp$^-$ mutants of *Salmonella typhimurium* the mutation affected the genes for the synthesis of one of the four enzymes of the biosynthetic pathway for tryptophan each of the mutant groups could be assumed to lack one of the four enzymes required for total biosynthesis of tryptophan, as shown in [55].

[55]

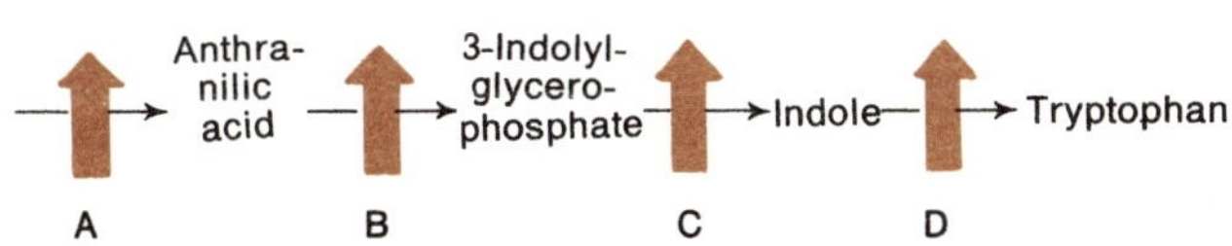

Accumulation of intermediates in cells or in the medium during cultivation of auxotrophic mutants can be used in industrial microbiology. This is the principle underlying the fermentation production of some amino acids (L-lysine and others) and some nucleic acid precursors. Auxotrophic bacterial mutants are at the basis of a modern branch of the fermentation industry, the products of which are beginning to find a place in the food, drug and chemical industries, as well as in the nutrition of farm animals.

An important group of mutants are those resistant to antibiotics or antimetabolites. Such resistant mutants can be isolated on solid media containing a sufficiently high level of the appropriate inhibitor. The technique thus makes it possible to inoculate media with a large number of cells as the selection medium employed permits only resistant variants to grow.

Some mutants which are resistant to antimetabolites possess an altered regulatory mechanism for a given metabolic pathway. Owing to this change, only a certain amount of the metabolite can be produced in the medium. Thus, for instance, mutants with derepressed histidine synthesis were isolated on the basis of resistance to 1,2,4-triazole-3-alanine and 3-amino-1,2,4-triazole. Such mutants can excrete histidine to the medium.

Mutants with a constitutive synthesis of a given enzyme (constitutive mutants) are isolated in a medium containing a co-repressor (p. 73). Mutants of *Escherichia coli* constitutive for alkaline phosphatase were isolated on a medium with β-glycerophosphate as the sole source of carbon and with inorganic phosphate as co-repressor, suppressing the synthesis of alkaline phosphatase in the inducible parent strain.

Phage-resistant mutants are usually distinguished by a loss of specific phage receptors. They are commonly isolated on solidified media in the presence of the specific phage.

Mutations modifying membrane permeability result in transport mutants. An example of such mutants is the *Corynebacterium glutamicum* strain, producer of glutamic acid. This strain does not synthesize biotin and hence biotin cannot participate in the synthesis of unsaturated fatty acids, which results in permeability changes of the membrane so that glutamic acid can be released into the medium. The same strain was used to derive mutants producing other amino acids, such as lysine, alanine and homoserine plus ornithine.

A simple selection technique based on cultivation at different temperatures permits one to obtain temperature-sensitive mutants which require either a higher or a lower temperature for growth than the parent strain does.

The phenotype of auxotrophic mutants is described according to the metabolite which they cannot synthesize or utilize, by attaching a symbol for the metabolite and a superscript minus sign; thus Urd^- is an auxotrophic strain unable to synthesize uridine, Lac^- is another strain unable to utilize lactose. Resistance of a mutant to streptomycin is designated with Sm^r and sensitivity to the antibiotic with Sm^s.

Colinearity between the Gene Structure and the Corresponding Protein

It is known from previous sections that the genetic information encoded in DNA is transmitted through replication to filial generations, transcribed into RNA structure and translated into proteins. The complex organization of the long peptide chain in a protein, an enzyme, must have its correlate in a segment of chromosomal DNA of the cell where the enzyme is synthesized. If a point mutation occurs in this segment of DNA the genetic code of the triplet in which the mutation has taken place is only rarely unchanged. Hence a different amino acid can be incorporated into the polypeptide [56].

[56]

Segment of parent strain	GAG CTC	GTT CAA	CCT GGA	AAA TTT	CCT GGA	TAA ATT	AGC TCG	CGG GCC
Colinear part of protein produced by parent strain	Leu .	Gln .	Gly .	Phe .	Gly .	Ile .	Ser .	Ala
Analogous DNA segment with two point mutations	GAG CTC	GTT CAA	CTT GAA	AAA TTT	CCT GGA	TAA ATT	ACC TGG	CGG GCC
Colinear part of protein produced by mutant strain	Leu .	Gln .	Glu .	Phe .	Gly .	Ile .	Trp .	Ala

Genetic map

Replacement of the TA pair for the CG pair results in a substitution of glutamic acid for glycine and a mutational change of the CG pair can cause replacement of serine by tryptophan in the resulting protein.

Testing colinearity between the sequence of base triplets in DNA and the amino acid sequence in the corresponding polypeptide is one of the outcomes of the molecular genetics of microorganisms. Mutants are usually applied, each of which possesses at least one mutational change in the gene investigated but at different sites (Fig. 36). The frequency of intragenic recombinants with no mutational change in the investigated gene is determined by the distance of the mutation sites on genes of crossed mutants (the notches in Fig. 36 represent point mutations on the DNA strand).

Another technique is based on the use of a marker gene during crossing, the gene being localized on the chromosome outside the region of the mapped locus. The marker in question is again a mutational change. The recombinant either contains the marker or it lacks it, and this indicates which of the two mutational sites under investigation lies closer to the marker on the genetic map (Fig. 37).

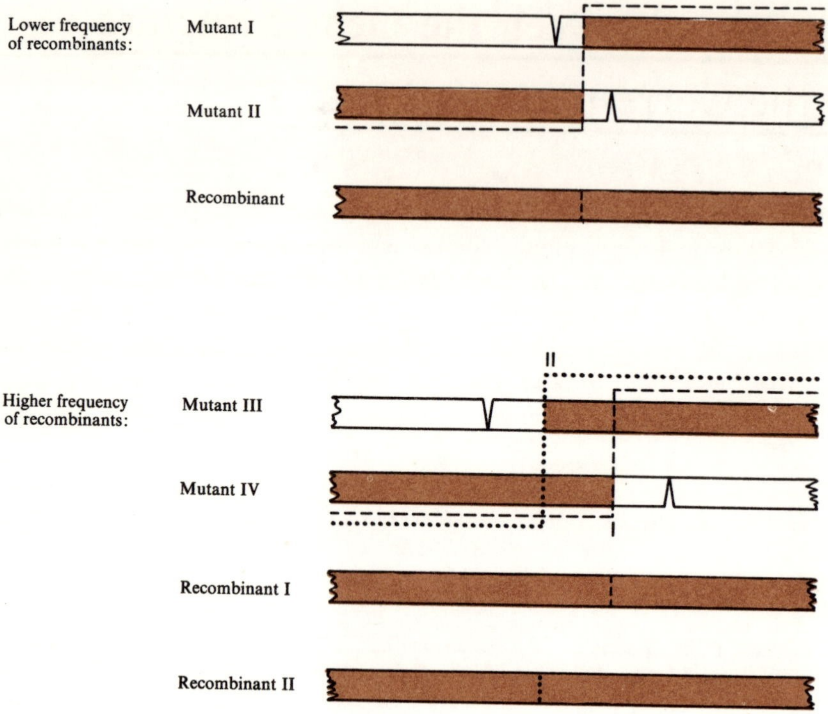

Fig. 36. Dependence of recombinant frequency on the distance of mutation sites

Fig. 37 shows that in the first case the sequence of mutation loci with respect to the marker was a – b – c while in the second case it was a – c – b.

Another technique for obtaining genetic maps is the use of deletion mutants in which it is examined whether the mutation site fits into deletions of the recombinant partner of different lengths.

The techniques can thus provide a very fine genetic map which can even distinguish between two mutants containing point mutations in two neighbour base pairs.

A fine piece of evidence of colinearity between DNA and the corresponding protein was provided by a study carried out at Stanford University. The protein studied was a subunit of tryptophan synthetase, a polypeptide containing 267 amino acids which is intensively synthesized in cells of *Escherichia coli.* The choice of this enzyme subunit was influenced by several favourable factors, mainly its high intracellular content and the existence of mutants which contained the polypeptide, but it was not catalytically active when it formed part of the enzyme molecule.

The task was to determine the primary structure (sequence of amino acids) of the polypeptide. The procedure was to split the polypeptide enzymically into smaller peptides, which were isolated and their amino acid sequence was determined. The overall sequence of the polypeptide was then put together from the analyses of the shorter peptides which overlap in many instances with other peptides and provide the key to the correct total sequence. The sequence of 267 amino acids was thus determined. The next part of the work was to obtain mutants of *Escherichia coli* in which the enzyme subunit was synthesized but which was inactive in the final molecule of tryptophan synthetase. The polypeptide subunits of the mutant strains were not analysed *in toto* but only the so-called fingerprints were examined: the polypeptide was again split (say, by trypsin) into a mixture of peptides, which were separated by paper chromatography in the first direction and by paper electrophoresis in the second direction. Ninhydrin detection will reveal a variety of spots differing in size, position and colour.

The fingerprints of the mutated subunits of tryptophan synthetase of *Escherichia coli* were all much alike, but in most instances a significant difference was observed: one of the peptides of the fingerprint was missing while another appeared which was not present in the parent strain. Both peptides, the missing one and the new one, were isolated and their exact sequence was determined. The peptides usually differed in a single amino acid and it was simple to determine the site at which the amino acid of the parent strain was replaced by another one in the mutant.

This site was then compared with the site of the presumed mutation change in the genetic map of the tryptophan synthetase subunit such as was obtained by genetic

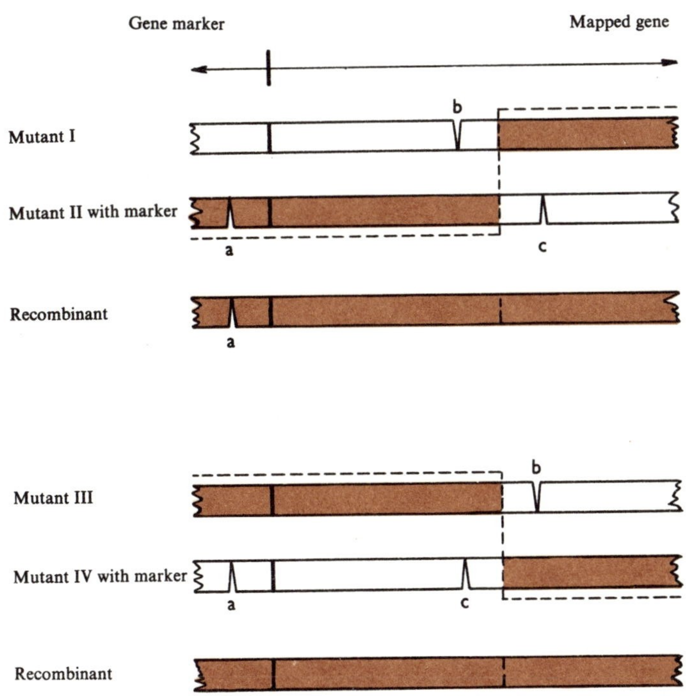

Fig. 37. Use of markers for determining the sequence of mutation sites

recombination of mutants of *Escherichia coli* using transduction with Plkc phage. Comparison of the mutation loci on the genetic map and the site of amino acid replacement then bore out the excellent agreement between the two. It was incidentally shown that two mutation loci lying almost at the same place on the genetic map resulted in the replacement of the same amino acid in the polypeptide: a point mutation took place in two base pairs lying in the same code triplet, the nearest points on the genetic map representing changes in neighbouring base pairs. Hence it follows that the number of possible mutation sites on a genetic map is equivalent to the number of base pairs in the gene.

Colinearity between the triplet sequence of DNA and the sequence of amino acids was successfully demonstrated in a number of biological models. Its existence is proof of the mechanism of translation and transcription of genetic information, as well as of the triplet nature of the genetic code and correctness of the molecular theory of point mutations.

Biological Movement of Genetic Material

Introduction

The temporal continuity of living organisms is based on their reproductive ability. Biological continuity between parents and their offspring thus necessarily rests on genetic continuity.

Hereditary properties, including the morphology and physiology of organisms, are known to be controlled by informational macromolecules capable of autoreplication. Replicated informational macromolecules are transmitted to the progeny in two ways, sexually and vegetatively.

1. During sexual reproduction the progeny are formed through fusion of two sexually distinct cells, the so-called gametes, which carry only half the number of chromosomes of somatic parent cells. Fusion of the male and female gametes results in a diploid zygote (fertilized mother cell) from which a multicellular organism can be formed. All its somatic cells have the same number of chromosomes as the somatic cells of the parent. Sexual reproduction is accompanied by genetic recombination: genes in which the parent organisms differed are joined here in new combinations.

2. During asexual reproduction the cell cycle is not one of alternating diploid and haploid phases. The new individual is not formed by fusion of two haploid cells. The daughter cells are formed from somatic cells or their groups. A case in point is the formation of a population from tubers or rhizomes when the daughter cells have genetic equipment identical with that of the original individual. The progeny thus formed is called a clone or vegetative clone progeny.

Classical genetics was studied in higher organisms which multiply sexually. The fundamental method of classical genetics was intentional crossing of individuals belonging to the same species but differing in at least one hereditary character and thus carrying different genotypes. This is the technique called hybridization. Examination and statistical evaluation of the progeny phenotypes are then the basis of genetic analysis.

The use of bacteria and viruses as model material made it possible to study hereditary characteristics on a molecular level and laid the basis for molecular genetics. In bacteria and viruses it is possible to investigate the role of a gene using the genetic analysis of hybrids. However, using special techniques, the genetically active material can be studied at the level of individual nucleotide pairs.

Bacteria and viruses offer the exceptional possibility of recombining the hereditary properties. The principle of such "hybridization" lies in transferring a part of the genetic material into a recipient cell where it can recombine with the corresponding section of the recipient cell genome.

In phages recombination takes place when genetically distinct mutants of the same phage type reproduce in the host cell. In bacteria three basic types of recombination processes are known: transformation, conjugation and transduction. A special type of cell is part of the process: a merozygote which contains the complete genome of the recipient but only a part of the genome of the donor. The recently introduced techniques of genetic engineering permit a highly rationalized DNA recombination, the often discussed gene manipulation.

Reduplication and Distribution of Chromosomes in Eukaryotes

Localization of the basic genetic information in the nuclear chromosomes of higher organisms elevates the nucleus to the supreme position in the hierarchy of cell organelles*. During cell division (cytokinesis) the nuclear material multiplies and redistributes itself by processes of fundamental importance.

Reduplication of chromosomes results in the formation of precise copies of the original chromosomes which are then transmitted to the daughter cells. Cytology permits the detection only of the phase when these equivalents are being separated. Optical observation of chromosome distribution is limited to dividing nuclei of some higher organisms with a well differentiated nucleus. Most somatic cells are diploid, containing two sets of chromosomes (2n). Only exceptionally does one encounter triploid (3n), tetraploid (4n) and polyploid cells. It is thus a simple multiple of the haploid number (n) which, however, can differ over a wide range among different organisms (p. 80). The haploid number of chromosomes is only found in the gametes of higher organisms.

Nuclear division (karyokinesis) which precedes cytokinesis may be of two types. Division which ensures the same number of chromosomes in the daughter and in the mother cells is called mitosis. Reductive division which decreases the original

---

* It was possible to transplant nuclei of amphibian somatic cells into enucleated eggs. During the first phase of the transplantation the chromosome-containing egg protoplasm is removed. This operation can be performed with a needle under microscopic control. In the second phase a nucleus is carefully removed from an isolated somatic cell. In the third, technically most demanding, phase the isolated nucleus is inserted into the enucleated nonfertilized egg. If the operation was performed with the nucleus of a tadpole intestinal cell and an enucleated frog egg normal frogs developed in several instances. This indicates that the nucleus of a specialized cell still contains all the genes required for the development of a complete animal. Specialization thus does not cause the loss of any genes although many will not be expressed in that particular cell. Transplantation of nuclei is of particular importance in studying tissue differentiation. Such transplantations have recently been used for cloning mice and sheep.

number of chromosomes to half is called meiosis; this type of division precedes the formation of gametes.

The time interval in which DNA is being reduplicated is an important section of the cell cycle, which can be divided into four phases of unequal length: M, the phase of mitosis; G1, the phase prior to DNA synthesis; S, the phase of DNA synthesis; and G2, the phase between DNA synthesis and mitosis. A part of the cell cycle is then the cytokinesis. Cell and nuclear division are controlled by different regulatory mechanisms.

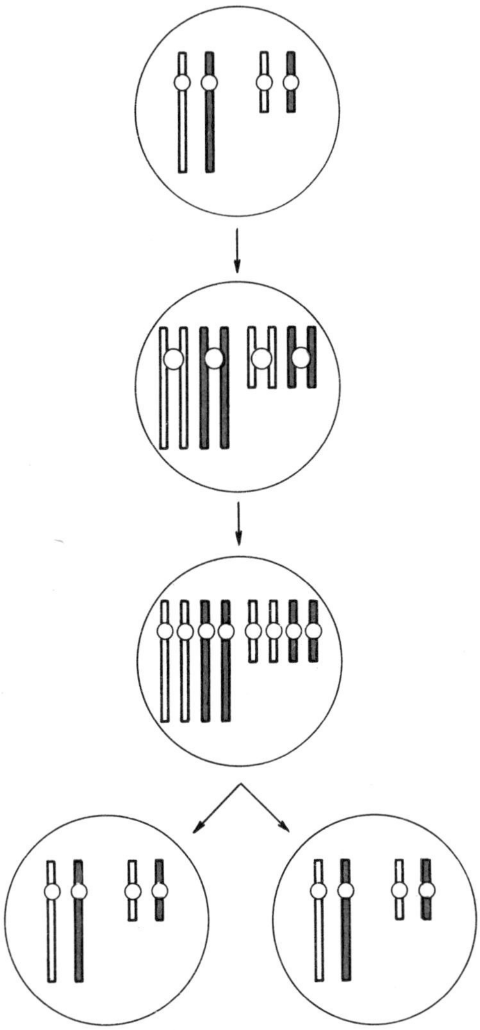

The nucleus of a diploid cell contains two pairs of homologous chromosomes before reduplication. Every chromosome is formed by one chromatid. The circles represent centromeres (in fact these are optically observable sites of constriction on the chromosomes).

Chromosomes were reduplicated during the interphase so that every chromosome now consists of two identical sister chromatids joined by the centromere.

At the end of metaphase the sister chromatids separate at the centromere.

Mitotic division yielded two daughter nuclei identical with the nucleus of the parent cell.

Fig. 38. Simplified scheme of mitotic nuclear division

Mitosis

Before a mother cell can divide into two analogous daughter cells, two processes must be completed: reduplication of the original chromosomes and distribution of the reduplicated chromosomes in such a way that the daughter cell contains complete and genotypically equivalent sets of chromosomes.

The diagram in Fig. 38 represents schematically the mitotic process with stylized chromosomes; it only expresses the mechanics of reduplication and distribution of the nuclear material. The morphological changes of the nucleus and the chromosomes as well as their movement during mitosis are shown in Fig. 39; the individual drawings are accompanied by a synchronized description.

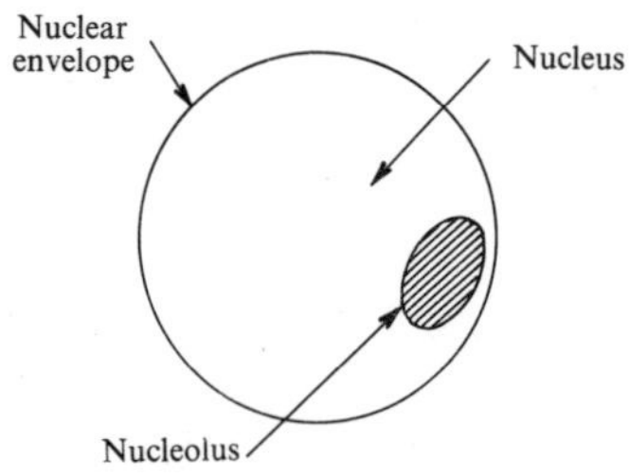

Chromosomes are first undetectable as specific forms. DNA is localized in the nucleus in an amount corresponding to two sets of simple chromosomes (every chromosome now corresponds to one chromatid). These uncoiled chromosomes reduplicate so that the amount of DNA in the nucleus is doubled (every chromosome now corresponds to two identical chromatids). This period between two mitoses is called the interphase. The actual stimulus for karyokinesis is not known.

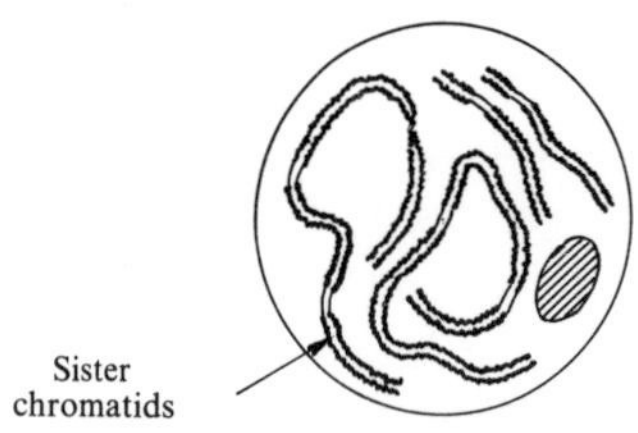

At the beginning of mitosis chromosomes begin to condense, coil and are reduced in length, their diameter increases and they become visible. During this period, the prophase, every chromosome may be seen to be composed of two sister chromatids joined by the centromere.
When the chromosome condensation reaches a certain level the nuclear envelope disappears and the mitotic spindle becomes visible. This spindle is responsible for correct separation of sister chromatids.

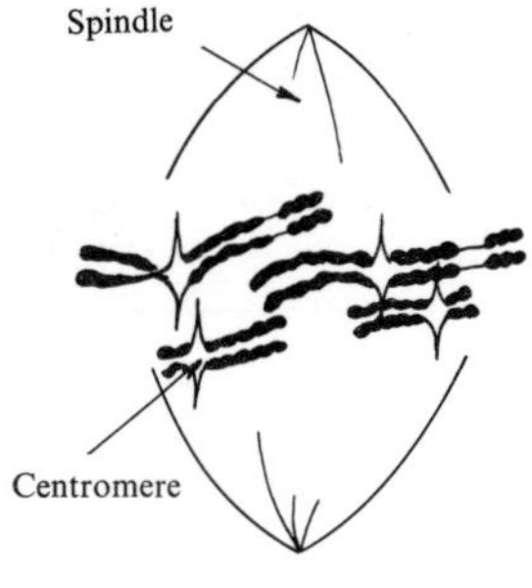

Condensation of chromosomes continues.
Chromosomes are now grouped equatorially, perpendicularly to the spindle axis. The spindle fibers connect with the centromere of every chromosome.
This is the metaphase. It is now that it is possible to determine the number of chromosomes and to identify homologous chromosomes by their shape so as to construct a cell idiogram (an artificial arrangement of images of all chromosomes according to homologous pairs and in descending order of sizes).

Sister chromatids separate at the centromere and begin to migrate toward the opposite poles of the spindle. This is the anaphase.

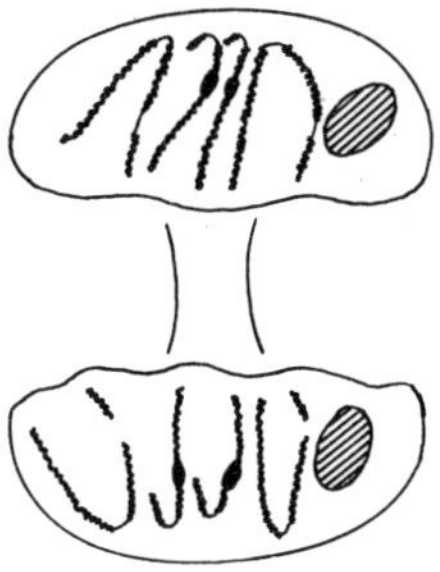

Chromosomes, now grouped at the opposite poles, uncoil and lose their distinct shape. This is the telophase.
At the same time, the spindle disappears and a nuclear envelope is formed which then encloses each of the separated sets of chromosomes.
Chromosomes are now fully uncoiled and the nucleus proceeds toward a new interphase. Nuclear division is followed by cytoplasmic division.

Fig. 39. Scheme of reduplication of nuclear material and its distribution into daughter cells during mitosis

Meiosis

The meiotic nuclear division takes place in higher organisms in the process of formation of sexual cells. The formation of such gametes is preceded by (a) reduplication of chromosomes of the diploid cell from which the gametes are differentiated, (b) reduction in the number of chromosomes to half (haploid) and (c) separation of the reduplicated sister chromatids (chromosomes) into the four future haploid nuclei.

The meiotic process is controlled genetically in both space and time and it is known to have several variants. Morphological changes of nuclear material during meiosis, as observed microscopically, proceed through several stages (Fig. 40); it is outside the scope of this book to discuss all the details but without a sound knowledge of the principles of meiosis it would hardly be possible to comprehend the fundamental laws of heredity of higher organisms.

Meiosis or the reductive nuclear division includes two subsequent divisions of the chromosome material. The first division is called heterotypic and results in a reduction of the chromosome number to half; the second division is homoeotypic and represents an analogy to the mitotic division. Each of the nuclear divisions proceeds through the four basic phases known from mitosis: prophase, metaphase, anaphase and telophase.

A diagrammatic illustration of the changes and distribution of chromosomes during meiosis is shown in Fig. 41.

The prophase of heterotypic division is genetically most important and its progress can be divided morphologically into five further developmental stages.

Chromosomes become visible under the microscope as long, uncoiled fibers not yet duplicated; this is the leptogenic stage of meiotic prophase which was preceded by DNA reduplication.

In the zygotenic stage of meiotic prophase homologous chromosomes form pairs (synapse of homologous chromosomes) and form the so-called bivalents. At the same time, internal coiling of the chromosomes results in their shortening and thickening so that they are clearly visible. The degree of coiling is not uniform along the entire length of the chromosomes; places with a higher condensation form nodes, called chromomeres. The number and sequence of these chromomeres are specific. Internal coiling proceeds throughout the heterotypic division.

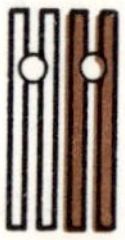

Bivalent

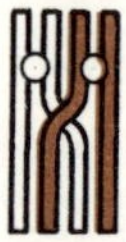

Chiasma

Recombination by crossing-over

In the pachytene stage of prophase of heterotypic division crossing-over can take place, a genetically important event. It permits recombination of linked genes, localized in homologous chromosomes. The chromosomes are clearly visible as bivalents. At the same time, sister chromatids differentiate and are joined two and two by centromeres. The bivalent has now four strands, two and two of which are nonsister. At this stage the crossing-over process takes place. The nonsister chromatids break at the same site and rejoin under reciprocal exchange of equivalent DNA segments. During the nonsister exchange the chromatids cross over. As soon as repulsive forces begin to act between homologous chromosome chiasmata are formed at the sites of crossing over which becomes particularly clear during the diplotene stage of meiotic prophase. The crossing-over thus accounts for intrachromosomal genetic recombination.

The diplotene stage continues by separation of homologous chromosomes and the remaining chiasmata are shifted toward chromosomal ends.

Diakinesis marks the end of the prophase of heterotypic division. Chromosomes are still internally coiled and appear as short and thick forms. The nuclear envelope still exists.

The spindle is formed along with the loss of nuclear envelope. The bivalents (synaptic pairs of homologous chromosomes) are oriented equatorially, perpendicularly to the spindle axis. This is metaphase I.

In the following anaphase I one chromosome of each pair moves toward the opposite pole of the spindle. Sister chromatids are still connected by the centromere. Selection of the direction movement of individual chromosomes is random and is thus a probability event. This makes interchromosomal gene recombination possible when genes localized in nonhomologous chromosomes are combined.

Heterotypic division is terminated by telophase I; in this stage the chromosomes are spatially distributed into two haploid sets but each is still composed of two chromatids.

This stage is followed by interkinesis when the chromosomes elongate again and partly uncoil.

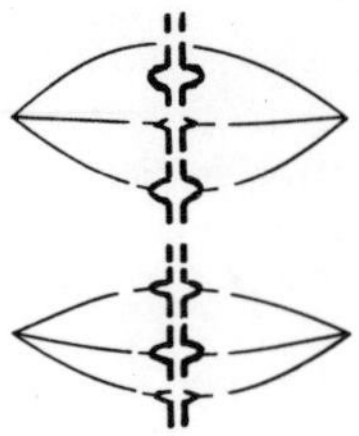

This is followed by the second, homeotypic, division which is analogous to mitosis.
During metaphase II the chromosomes are grouped in the equatorial places of the spindles and centromeres reduplicate.

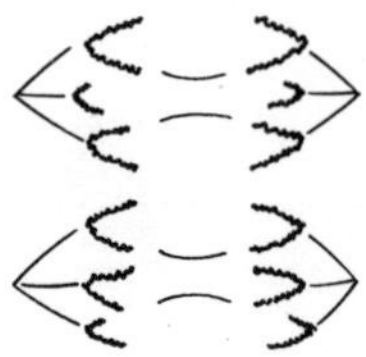

During anaphase II the sister chromatids wander toward the opposite poles of the spindle.

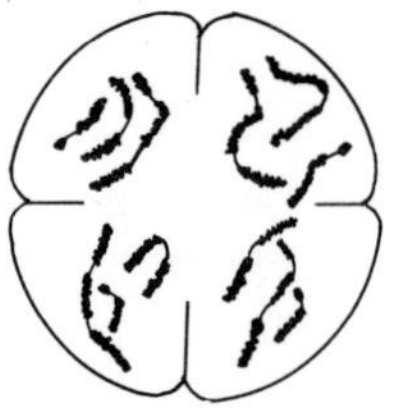

During telophase II four haploid nuclei are formed and these represent the final products of meiotic nuclear division.

Fig. 40. Simplified scheme of meiotic nuclear division

A special type of chromosome are the giant or polytenic chromosomes, which are of special importance in cytogenetic studies. The giant chromosomes may occur in the salivary glands of insect larvae (particularly the *Diptera*) and are generated by a multiple repetition of the reduplication cycle of the individual chromatids. The number of filaments of some giant chromosomes may be 2 000 and the chromosome may attain a size 200 times greater than that of the corresponding chromosome in a normal somatic cell. The giant chromosomes contain a large number of longitudinally placed chromonemata (the smallest chromosomal filaments still seen in an optical microscope). Chromomeres of the polytene chromosomes are visible as compact discs. The position and size of the discs are the same in organisms of the same genotype. If the genotype is altered (e.g., after X-irradiation), certain structural differences may be observed even in the giant chromosomes. The changes can be directly followed optically and the position of genes for certain characteristics may thus be directly established. The estimated number of genes (more than 5 000 in *Drosophila*) corresponds to the number of discs in the giant chromosomes. As an example of this type of chromosome, Fig. 42 shows the polytene chromosome of *Chironomus*.

As one may observe cytologically the activation of a certain site in polytene chromosomes, the model is often used for studying regulations; the site in which mRNA is synthesized can be seen here as a disc of lower density (the so-called puffs).

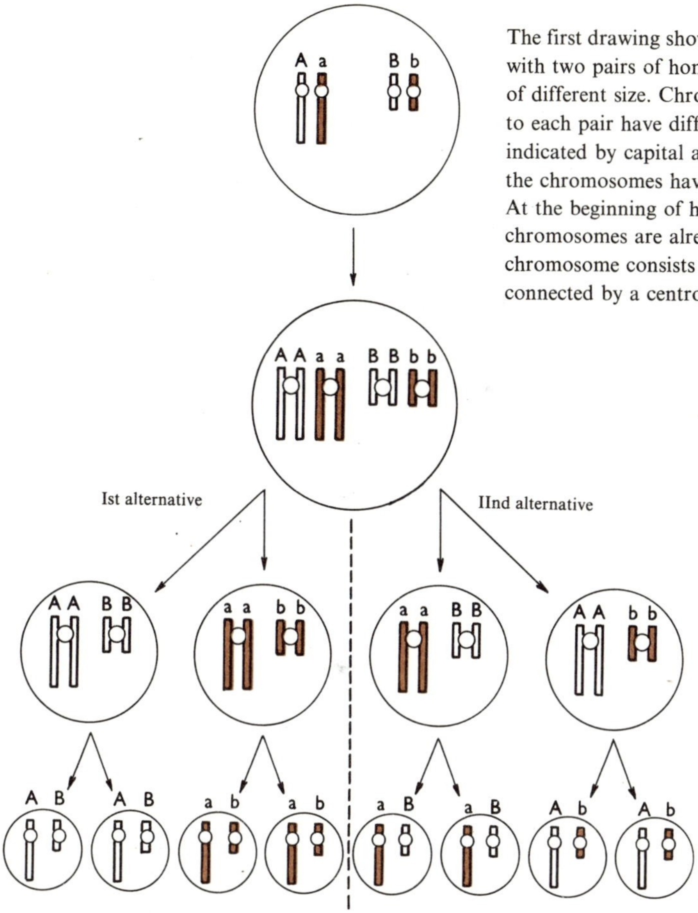

The first drawing shows the nucleus of a diploid cell with two pairs of homologous chromosomes of different size. Chromosomes belonging to each pair have different types of genome as indicated by capital and small letters. At this stage the chromosomes have not reduplicated.
At the beginning of heterotypic meiotic division chromosomes are already reduplicated. Every chromosome consists of two sister chromatids connected by a centromere.

At the end of heterotypic division the chromosomes divide into two sets with a haploid chromosome number. Sister chromatids remain connected by the centromere.
(For simplicity, the scheme does not show the crossing-over process.)
The randomness of the direction of movement of the individual chromosomes of each homologous pair is shown by alternatives I and II.
Homeotypic division results in four haploid nuclei in which every chromosome is represented by one chromatid.

Fig. 41. Scheme of reduplication of nuclear material and its distribution during meiosis

Reduplication and Transfer of Chromosomes in Bacteria

During exponential growth the bacterial cell may contain, depending on the rate of growth and the composition of the nutrient medium, between one and four chromosomes (genophores). These chromosomes have the same genetic structure and are derived from the same parent chromosome. In principle a bacterial cell is thus haploid. The bacterial chromosome has no nuclear envelope and no spindle-like device can be observed during division and distribution of bacterial chromosomes.

Fig. 42. Polytene chromosome of *Chironomus*

Various hypotheses exist about the separation of bacterial chromosomes into daughter cells. An important role is ascribed here to the mesosome, an intracytoplasmic membrane formed by invagination of a part of the cytoplasmic membrane. The bacterial chromosome is bound to the mesosome and separation of reduplicated chromosomes is then accomplished by membrane growth. This membrane acts as both a temporally and locally coordinating element for DNA reduplication and for division of the mother cell. The mechanism is shown schematically in Fig. 43.

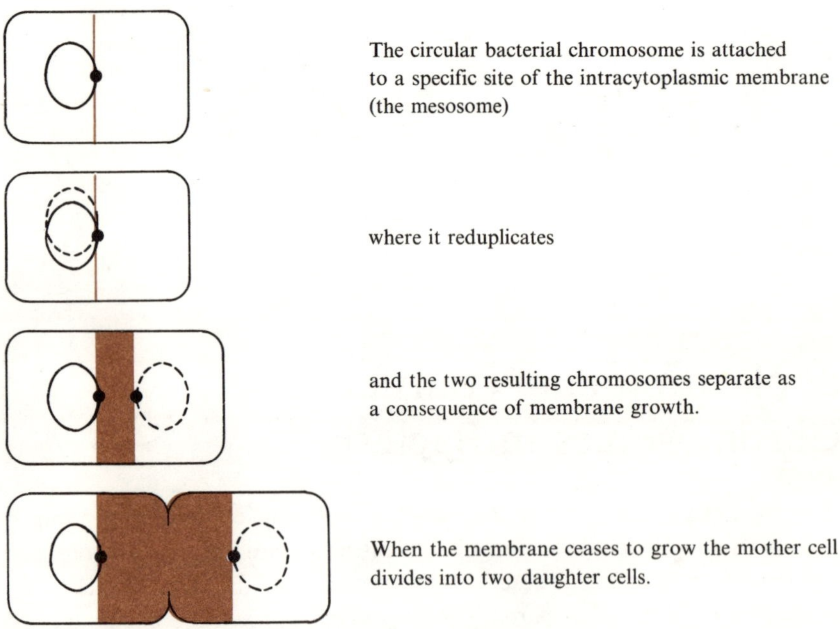

Fig. 43. Scheme of reduplication of the bacterial chromosome with closed structure and the mechanism of distribution of the chromosomes into daughter cells

Analysis of Fundamental Genetic Principles

Mendelian Laws

The results of Mendel's experiments with pea crossing were published in 1865. Mendel followed for several consecutive generations a number of well defined and genetically stable morphological characteristics of pea, such as colour and shape of the seed, shape of the pod, position of the flower on the stem and height of plants. He worked with a large number of experimental plants and observed that the inheritance of the characters follows certain rules. Mendel's laws were "rediscovered" at the beginning of this century and re-evaluated by Correns, Tschermak and de Vries.

The simplest case of hybridization may be represented by a parent plant of pea (the parental, or P, generation) which is crossed and its offspring studied for several generations. We shall follow here a single characteristic trait in which the two parents differ, say seed colour. The seed colour of pea is genotypically determined. One of the parents will constantly produce yellow seeds and the other green seeds.

After mutual pollination of the parent plants yellow seeds will be produced, the colour being that of the cotyledons from which the next-generation plant will evolve. These seeds then give rise to the first filial (F1) generation. After mutual pollination of the F_1 plants, seeds with embryos of the second filial (F_2) generation will be formed. The colour of these seeds is not uniform: there will be both yellow and green seeds, in a ratio of 3:1. Hence it will be observed that the seeds of the F_1 generation will be of uniform colour no matter which of the parents was the pollen donor. The characteristic which predominated was called dominant by Mendel, the one that was suppressed was called recessive.

Let us now undertake a genetic analysis of the above hybridization experiment. The diploid cells of both parents contain hereditary determinants (genes), responsible for seed colour, in two forms or alleles. They are localized at analogous loci of homologous chromosomes. Alleles of the dominant type, such as those responsible for the yellow colour of pea seeds, are designated with capital letters (**AA**). The designation of alleles usually stems from the initial letter of the coded "property" in Latin or in English. Recessive-type alleles, such as those responsible for the green colour of seeds, are designated with the same but lower-case letters (**aa**).

If the two alleles for the given property are identical we are dealing with a homozygous state (**AA** or **aa**); if they are not identical the state is designated as heterozygous (**Aa**).

The allele pairs are segregated independently of each other as self-contained units during gamete formation (see the discussion of meiosis) and can then be combined in different ways during fertilization.

The formation of the first filial generation was thus preceded by gamete formation, each of the gametes containing a single allele for seed colour. The allele for seed colour in one parent is only dominant (**A**) and in that of the other parent it is only recessive (**a**). The hybrid of the first filial generation must then be heterozygous with respect to the genotype for seed colour; its diploid somatic cells contain a pair of alleles of type **Aa**. Owing to the presence of the dominant allele **A**, the phenotypic expression of this genotype is the yellow seed colour and these seeds will not differ in their appearance from seeds produced by the homozygous parent plant with alleles of type **AA***. The analysis is shown schematically in Fig. 44.

One of the fundamental principles of genetics of higher organisms is that of independent assortment of genes, also discovered by Mendel. The principle will now be demonstrated on the example of crossing two pea plants differing in two characteristics, colour and surface texture of seeds. Each of the characteristics is determined in diploid cells by a single pair of alleles. Alleles responsible for the yellow colour and smooth surface of seeds are dominant and those for green colour and wrinkled surface are recessive. The plant genotype where both pairs of alleles for these characteristics are dominant will be designated with **AABB** (**AA** for yellow colour, **BB** for smooth surface). The plant genotype where both alleles are recessive will be designated **aabb** (**aa** for green colour and **bb** for wrinkled surface). Crossing of these homozygous parent plants gives rise to the first filial (F_1) generation with a uniform phenotype and genotype **AaBb**. The presence of dominant alleles for the two markers determines the phenotype, in this instance the appearance of seeds that are yellow with a smooth surface. The second filial (F_2) generation arising from crossing the plants of the F_1 generation contains several different genotypes. The hybridization experiment is shown schematically in Fig. 45.

It will be seen that nine out of sixteen seeds are yellow and smooth, three are green and smooth, three are yellow and wrinkled and one is green and wrinkled. The phenotype segregation ratio is thus 9:3:3:1.

It follows further from the scheme that the genes under investigation can combine freely and independently, which is only possible if they are located in different pairs of homologous chromosomes. Interchromosomal recombination of individual alleles takes place during meiosis, which is known to precede the formation of gametes. During

---

* The relationship between dominance and recessivity is not the only functional connection between two alleles of a pair characterized by a certain phenotype. A situation may arise when a phenotype determined by two alleles of the same pair is intermediate between the two extremes, the so-called semidominance. A hypothetical experiment might illustrate this case. If two parent plants are crossed with homozygous alleles for blossom colour, one red and the other white, the progeny in the F_1 generation (with a heterozygous genotype) will have pink blossoms. The character is thus phenotypically intermediate. The F_2 generation will then contain individuals with red, pink and white blossoms at a ratio of 1 : 2 : 1.

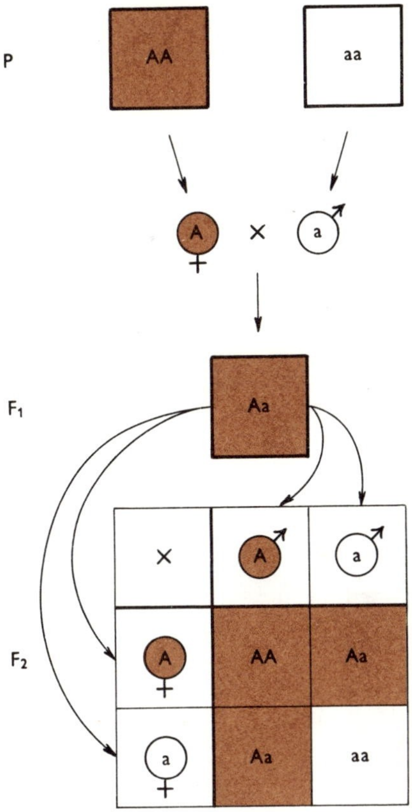

Fig. 44. Hybridization experiment to follow a single character.
P, Parental generation; F_1, first filial generation; F_2, second filial generation

Both alleles of the maternal plant in the parental generation are of type **A**, those of the paternal plant of type **a**.

Gametes of the maternal plant are only of type **A**, those of the paternal plant only of type **a**.

A pair of alleles of somatic plant cells formed after fertilization is always of type **Aa**. The gametes of these plants are of two types – either they carry allele **A** or they contain allele **a**, at a 1 : 1 ratio.

During fertilization the combinations shown in the Mendelian square are formed. Three of the possible combinations contain the dominant allele and yield yellow seeds while one of the four combinations lacks the dominant allele and will produce green seeds. The ratio of yellow to green seeds will be 3 : 1 (this is the phenotypic segregation ratio). However, genotypically three different categories are present: a dominant homozygote, a heterozygote, and a recessive homozygote, at a ratio of 1 : 2 : 1.

meiotic division the homologous pairs of chromosomes are segregated into two haploid sets. It is a random decision in which of these two haploid sets a given chromosome from a homologous pair will appear. The number of possible combinations of these unpaired chromosomes (and hence transferred alleles) corresponds to the number of combinations between independent units, i.e., chromosomes.

Fusion of gametes formed during meiotic division then gives rise to different genotype combinations. The number of these combinations and their statistical representation in the F_2 generation depends on the number of characters investigated and follows mathematical rules (Table VIII).

On the basis of hybridization experiments, Mendel's laws may be formulated and they are now known to be generally valid:

The principle of gene segregation demonstrates that paired alleles (genes) separate during gamete formation as independent units.

The principle of independent assortment of genes indicates that during gamete formation independently combinable genes are only those located on different pairs of chromosomes.

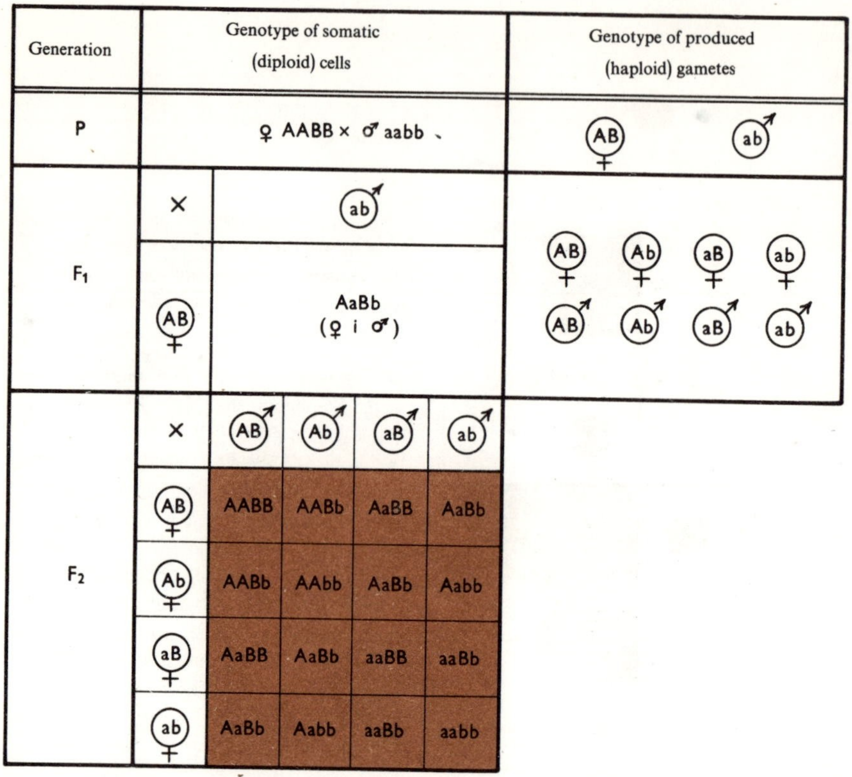

Fig. 45. Hybridization experiment to follow two freely combinable characters. P, Parental generation; F_1, first filial generation; F_2, second filial generation

Table VIII. Quantitative analysis of mono-, di-, tri- and n-hybrid crosses under complete dominance and independent assortment of genes

Number of characters examined	Number of gamete types with different genotypes	Number of gamete combinations in F_2 generation	Number of different genotypes in F_2 generation	Number of different phenotypes in F_2 generation	Frequency of individual phenotypes in F_2 generation
1	$2^1 = 2$	$4^1 = 4$	$3^1 = 3$	$2^1 = 2$	$(3 + 1)^1 = 3 + 1$
2	$2^2 = 4$	$4^2 = 16$	$3^2 = 9$	$2^2 = 4$	$(3 + 1)^2 = 9 + 3 + 3 + 1$
3	$2^3 = 8$	$4^3 = 64$	$3^3 = 27$	$2^3 = 8$	$(3 + 1)^3 = 27 + 9 + 9 + 9 + 3 + 3 + 3 + 1$
n	2^n	4^n	3^n	2^n	$(3 + 1)^n$

Recombination of Linked Genes

The number of genes determining the individual characteristics of an organism may attain thousands even in such simple objects as bacteria. The number of chromosomes is smaller by several orders of magnitude. A bacterial cell even contains a single chromosome (genophore) even if occasionally in several copies. It is thus obvious that every chromosome must contain a great number of genes. Hereditary association of some genes based on their localization in the same chromosome or the same molecule of DNA is called gene linkage. As chromosomes are redistributed during reproduction as discrete units, genes present in a chromosome appear as mutually linked.

In diploid organisms every character is most simply determined by two alleles (a single pair of genes) which are located in analogous positions on a pair of homologous chromosomes. Every pair of homologous chromosomes thus forms a linkage group of genes. The number of linkage groups is therefore identical with the haploid number of chromosomes.

Complete (absolute) linkage was in fact observed in rare cases, such as sex-linked hereditary characters. Linkage is usually incomplete and linked genes may be segregated and rearranged in new genotype combinations. We are dealing in such cases with recombination of linked genes or with the so-called intrachromosomal recombination, based on the crossing-over process.

Recombination of linked genes in eukaryotic cells usually takes place during meiosis (meiotic recombination). In special cases, such as with some fungi, this type of recombination may even occur during mitosis; this is then mitotic or somatic recombination (somatic crossing-over). Meiotic and mitotic intrachromosomal recombinations are reciprocal in most instances. Certain segments of genetically active material are substituted for homologous segments which may differ in quality. For recombination to occur, this mutual substitution must take place between non-sister chromatids of paired chromosomes. Intrachromosomal recombinations may even take place in the bacterial chromosome with a circular DNA strand during recombination processes, such as transformation, conjugation and transduction.

Recombination studies are at the basis of genetic mapping. From an analysis of the frequencies of different recombinants in a large population it is possible to determine not only the distances between individual linked genes but also the sequence of individual genes. Gene linkage was first described in 1906 by Bateson and Punnett.

Gene Linkage in Eukaryotes

It was briefly explained above that the principle of gene linkage in diploid organisms consists of the fact that a pair of chromosomes contains genetically active material which determines a number of different characteristics. The genetic consequences of gene linkage are best demonstrated by an example.

Let us examine the results of a hybridization experiment in which two characters will be followed, each of them determined by a single pair of alleles; both genes are positioned on the same pair of homologous chromosomes. Fertilization is accomplished between parents, one of which carries the dominant alleles for both characters (the genotype of the diploid cells is homozygously dominant **AABB**), while the other carries the recessive alleles for both characters (its genotype is homozygously recessive **aabb**). The F_1 generation will have a uniform phenotype which corresponds to the dominant homozygous parent type (**AaBb**). The F_2 generation, however, will contain genotypes (and consequently phenotypes) at ratios which do not agree with the theoretical predictions of Mendel's laws. The cause of this lies in the localization of both allele pairs in the same pair of homologous chromosomes. The situation arising during hybridization of linked genes is shown in Fig. 46, using stylized chromosomes carrying two allele pairs.

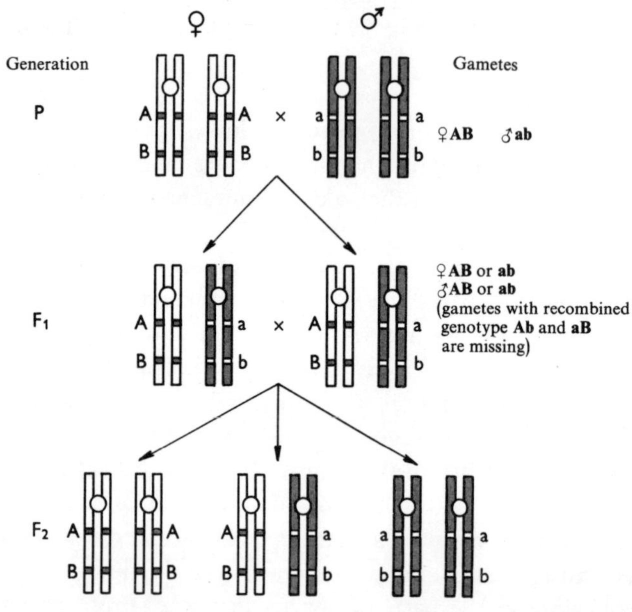

Fig. 46. Graphical representation of gene linkage during hybridization with no crossing-over. P, Parental generation; F_1, first filial generation; F_2, second filial generation

According to Mendel's laws, meiosis of independently assorted genes results in independent segregation of alleles and the dihybrid individual of the F_1 generation produces four genotypically distinct types of gametes in equal amounts (**AB**:**Ab**:**aB**:**ab** = 1:1:1:1). If the genes are linked, meiosis does not result in segregation of the individual alleles as they are positioned in the same chromosome. If no crossing-over then takes place between non-sister chromatids, the dihybrid individuals of the F_1 generation will produce only two genotypically distinct types of gametes. If some crossing-over takes place, gametes of a recombined type may appear. The probability that crossing-over will take place decreases with decreasing distance between these two genes on the chromosome. The theoretically possible number of gametes with a recombined genotype (i.e., **Ab** and **aB**) is 50% of the total as genes are recombined by the crossing-over process only between non-sister chromatids in the bivalent. Any mutual substitutions between sister chromatids cause no change of the genotype or of the phenotype.

Linkage Phases and Consequences of Crossing-Over

The above example of recombination of linked genes concerned the situation when one of the parents was homozygously dominant and the other homozygously recessive for both characters. Such an arrangement of allelic pairs is called the coupling linkage phase and the genotype is designated as **AB*/AB** or **ab/ab**. However, frequently each of the parents used for hybridization is homozygously dominant for only one character and homozygously recessive for the other character. This linkage phase is called repulsion and the genotype is abbreviated to **Ab/Ab** or **aB/aB**. The allelic sets of linked genes do not change during crossing unless crossing-over takes place during meiosis. Gametes with nonrecombined genotype are formed preferentially during meiotic division. Genotypes of gametes with recombined alleles are different, depending on the type of the linkage phase. In the coupling phase the gametes with recombined genotype have alleles **Ab** and **aB** while in the repulsion phase the alleles are **AB** and **ab.** Individuals arising from germ cells with recombined genes are called recombinants. Their gene linkage groups are different from those of their parents.

The consequences of crossing-over for a dihybrid with both the coupling and the repulsion linkage phases are shown schematically in Fig. 47, which indicates that the maximum number of gametes with recombinant genotype may be 50% as the reciprocal substitution is expressed phenotypically only between non-sister chromatids.

Linkage Relation and Ways of Determining It

The degree of gene linkage is determined in most instances by the distance between genes in the same chromosome, and hence it decreases with the distance between the positions of two alleles investigated. Genes (alleles) placed next to each other are linked very tightly and the linkage relation is maximal: genes (alleles) placed at the opposite ends of a chromosome have the lowest linkage strength.

---

* The fraction shows the sets of linked alleles determining the characters examined in diploid cells; the numerator will contain symbols for alleles localized in one of the chromosomes, and the denominator the corresponding alleles in the other chromosome of a pair.

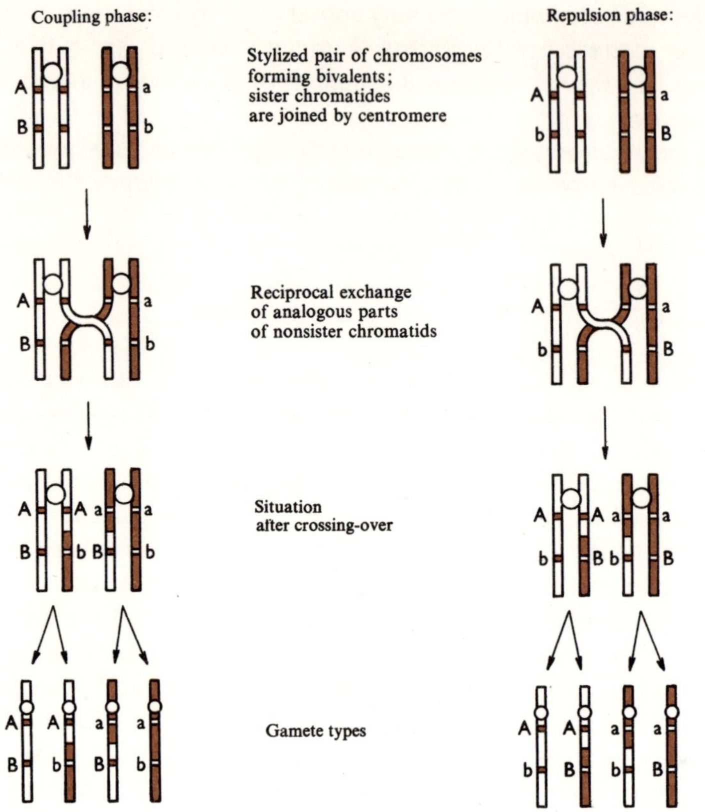

Fig. 47. Graphical representation of gene linkage during hybridization including crossing-over

The linkage relation is established from the genetic consequences of crossing-over processes. The probability of reciprocal recombination of linked alleles by crossing-over increases with increasing distance between the alleles on the chromosome. Hence, the degree of linkage can be calculated from the frequency of recombinants formed from the gametes.

The frequency of recombinants is easily determined by crossing a heterozygous dihybrid of the F_1 generation with an individual which is homozygously recessive in both characters. The resulting genotypic equipment of the crossed individuals now permits one to determine directly the frequency of production of recombined gametes. The setup and results of this hybridization experiment (for both coupling and repulsion linkage phases) are shown in Fig. 48.

The fraction of recombined phenotypes in the total of progeny of back crosses is called Morgan's number and is denoted by p; the unit of linkage is one morgan, which represents one per cent of recombinants. The p values vary from 0 to 0.5 (i.e. from 0 to 50%). If $p = 0$ the linkage is complete and there is no recombination after crossing. If $p = 0.5$ the genes are independently combinable. The degree of linkage thus increases with decreasing number of recombinants.

Chromosome Maps

It was shown above that the gene linkage relation is proportional to the closeness of their position on the chromosome. A schematic representation of the gene sequence based on the relative distances between loci is called the chromosome map. To map a chromosome and to determine the sequence of linked genes the so-called three-point tests are applied. They again involve a hybridization experiment in which the positions of three linked alleles are examined.

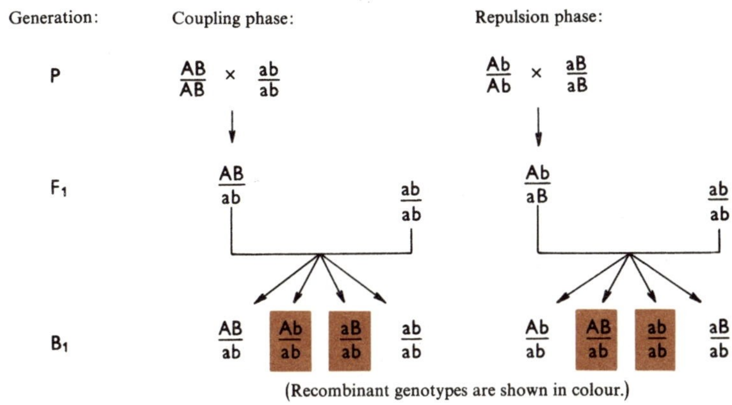

Fig. 48. Hybridization experiment involving gene linkage.
P, Parental generation; F_1, first filial generation; B_1, back crossing

A simplified example of such a three-point test will be presented in a hypothetical experiment in which two organisms with diploid chromosome numbers will be crossed. Each of the three linked characters is determined by a single pair of alleles; the dominant ones are again in capitals and the recessive ones are in lower-case letters. One of the parents is heterozygous with an **AbC/aBc** genotype and the other is homozygously recessive in all three characters, thus **abc/abc**. This recessive partner produces only **abc** gametes, which cannot influence the phenotype of the progeny. The progeny phenotype will be determined only by the genotype of gametes of the heterozygous partner which produces original nonrecombined parent-type gametes **AbC** and **aBc**. Gametes with a recombined gene structure will only be produced in cases when crossing-over took place during gamete differentiation (i.e., during meiotic division). If the crossing-over took place between genes **a** and **b**, the resulting gametes are **abC** and **ABc**. If the crossing-over took place between genes **b** and **c**, the resulting gametes are **aBC** and **Abc**.

The setup of the three-point test and its results are summarized in scheme [57]. The offspring contains six categories of phenotypes but, as follows from [57], the numbers of individuals in the categories are very different.

[57]

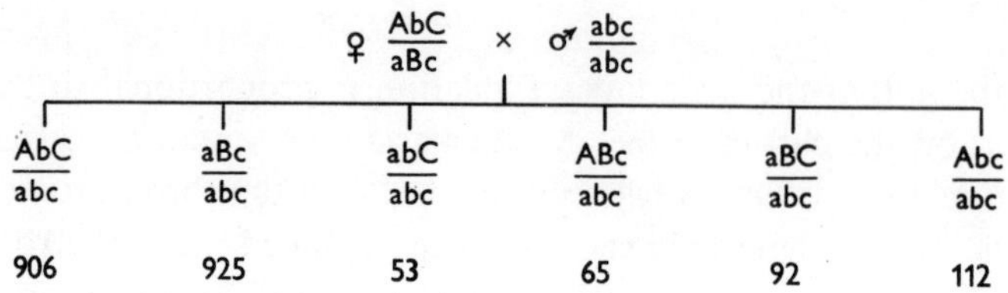

The phenotypic expression was thus evaluated in 2153 hybrid individuals. Of these, 1831 were of the parent (nonrecombined) type. There were 118 recombinants between **a** and **b** genes, i.e., 5.4%, 204 recombinants between **b** and **c** genes, i.e., 9.4%, and those between **a** and **c** represent the recombinant total, i.e. 322 or about 15%.

Nonrecombined phenotypes	**AbC**	906	1831
	aBc	925	
Recombined phenotypes after crossing-over between **a** and **b**	**abC**	53	118
	ABc	65	
Recombined phenotypes after crossing-over between **b** and **c**	**aBC**	92	204
	Abc	112	
Recombined phenotypes between **a** and **c** (equal to sum of recombinants between **a** and **b** and between **b** and **c**)	**abC**	53	322
	ABc	65	
	aBC	92	
	Abc	112	

The values permit an estimate to be made of the relative distances between the genes examined and of their sequence on the chromosome. As the number of recombinants increases with increasing distance between genes on the chromosome the distance between genes **a** and **b** is less than that between genes **b** and **c**, specifically 5.4 vs. 9.4. The total, represented by 15, is practically the sum of the two smaller values and hence the sequence of genes on the chromosome appears to be **a**−**b**−**c** [58].

[58]

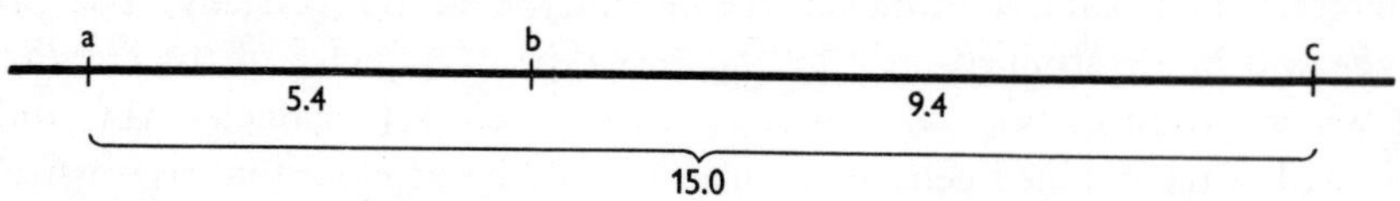

Recombination of genes **a** and **c** with position **b** preserved can come about only through a double crossing-over. The possibility of double crossing-over processes has been verified experimentally and even multiple crossing-over events are known.

The probability of a double crossing-over again decreases with decreasing distance between the genes studied and the theoretical number of double crossing-over events is expressed by the product of the two single crossing-over probabilities. In the **a** − **c** segment studied here it would then be 0.094 . 0.054, i.e., 0.507%. In the case described the result of a double crossing-over would be a combination of **abc** and **ABC***.

Morgan's Laws

On the basis of studies of gene linkage in *Drosophila*, Morgan prepared topographical genetic maps of its chromosomes. The maps clearly show a linear arrangement of genes in the chromosome (Morgan's first law). No case of branched organization of a chromosome has ever been detected. This linearity was directly demonstrated by changes of linkage between genes due to chromosomal aberrations.

Chromosomal aberrations are cytologically detectable changes in the number and sequence of genes carried by various chromosomal segments. A chromosomal aberration is usually accompanied by extensive structural changes of a chromosome and may even involve a change in the number of chromosomes. Such aberrations are caused by spontaneous or induced mutations.

The most common case of chromosomal aberration is a defect brought about by the loss of the terminal segment of a chromosome. Deletion is a damage caused by eliminating an internal segment of a chromosome. Inversion describes a turn of a chromosomal segment by 180°. Shifting of a chromosomal segment to another chromosome is called translocation. Chromosomal aberrations may occur in diploid organisms in both the homozygous and the heterozygous state, but homozygous aberrations are rare. The various types of chromosomal aberrations are depicted in Fig. 49.

Chromosomal aberrations must have played an important role in the evolution of species. A case in point is the number of chromosomes in apes ($n = 24$), which was reduced in man (*Homo sapiens*) to $n = 23$. The reduction is apparently caused by fusion of two smaller chromosomes to a larger one.

The first law of Morgan holds not only for the chromosomes of higher organisms but even for bacterial or viral chromosomal equivalents.

As mentioned before, the set of genes localized on a pair of homologous chromosomes of a diploid cell forms the linkage group. In other words, the two alleles determining a certain character are found in the same positions on the two homologous

---

* The actual occurrence of double crossing-over events is generally lower than predicted by theory. A role is played here by the so-called positive interference as a crossing-over event decreases the probability of another in the adjacent segment of the chromosome. An opposite phenomenon, one of negative interference, can occur between closely linked genes. Recombination then occurs at a higher than theoretical frequency. This phenomenon is often found during bacterial recombinations. Some segments of the bacterial chromosome are more prone to undergo recombination than others, apparently owing to nonuniform fragility of the DNA strand.

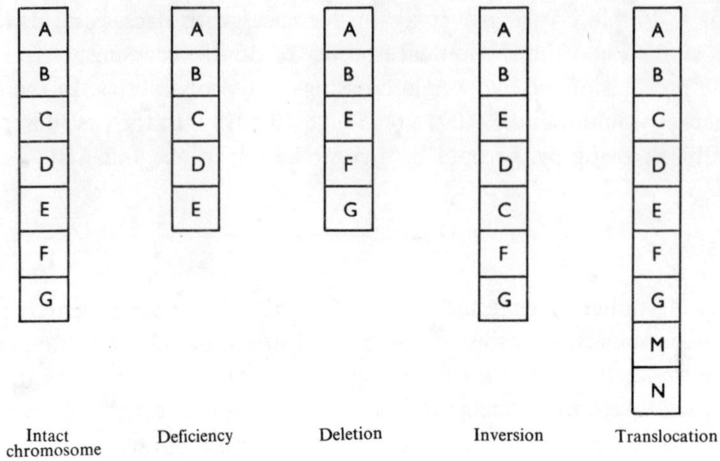

Fig. 49. Principal types of chromosomal aberrations

chromosomes. The number of linkage groups is thus equivalent to the number of pairs of homologous chromosomes (second law of Morgan). Thus, *Neurospora* contains seven chromosome pairs which can be directly observed in an ascus. Correspondingly, this ascomycete contains seven linkage groups. The determination of the number of linkage groups can aid in assessing the number of chromosomes in cases when cytological methods are not applicable (e.g., in some yeasts).

Recombination of Linked Genes in Bacteria

In higher organisms which multiply sexually, diploid (somatic) cells and haploid (sexual) nuclear phases alternate. In this way not only genetic factors but also new genotype systems are sure to be transmitted to the progeny. The process of crossing-over permits recombination of genes which are linked by being localized in the same chromosome. In a restricted sense recombination represents the generation of new sets of linked genes.

The term recombination often appears in bacterial genetics. Here, too, recombination leads to new sets of genes. As bacteria usually contain a single chromosomal equivalent (even if in several copies), individual genes appear as linked. Again, the degree of gene linkage is reflected in the frequency of recombinants. The best studied bacterial chromosome is the circular chromosome of *Escherichia coli* K12, representing a single linkage group. In bacterial recombinations effected by transformation or transduction only those genes are considered to be linked which are transmitted by the same fragment of DNA of the donor cell.

The mechanisms of genetic transfer, including transformation, conjugation and transduction, always represent a unidirectional transport of a part of the genetic material from the donor cell to the recipient cell. The result of this process is a partly diploid zygote (also called merozygote or heterogenote).

Genetic material received from the donor cell cannot be maintained permanently in the recipient cell as an independently multiplying unit. The principle of gene recombination in bacteria consists in an exchange of a genetically active segment of donor DNA and a corresponding homologous region of the chromosome of the recipient cell in the course of existence of a partly diploid state. However, this transient state may persist for several generations. The genetic recombinant thus formed then contains some of the original genome (genome of the recipient cell) and some of the genome of the donor cell. One of the possible mechanisms of the recombination process is depicted in Fig. 50.

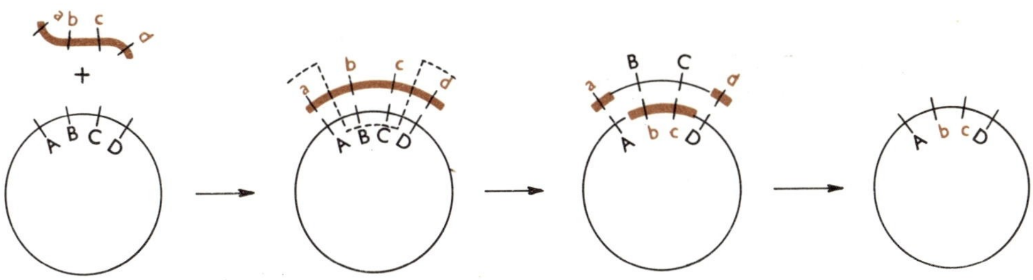

Fig. 50. Mechanism of exchange of donor and acceptor DNA during recombination

Theoretical Aspects of Recombination Mechanisms

The views about the mechanisms of crossing-over acquired a more concrete form only after it had been established that the gene is nothing but a segment of DNA. The first of the recombination theories assumes that during the crossing-over event breaking and rejoining of DNA molecules takes place at exactly corresponding sites of the recombining chromatids (Fig. 51).

Interruption of the two recombining strands at exactly the same site is not involved in the second theory, which is called the copy-choice mechanism (Fig. 52). According to this theory, the replication of paired chromatids is started by reduplication of the "paternal" strain of DNA and at a given moment the copying machinery switches to the "maternal" (non-sister) strand. If the nascent strand copying the "maternal" strand is shifted to the "paternal" one at the same site this will give rise to two reciprocally recombined strands.

In tetrad analysis of ascomycetes (see the section on the genetics of fungi), some irregular segregation ratios were found for linked genes, indicative of a non-reciprocal exchange between non-sister chromatids. In higher organisms a non-reciprocal recombination is very rare and occurs in genes with a high degree of linkage. An explanation of the mechanism of non-reciprocal recombination is offered by the theory according to which recombination proceeds via copy choice; in a certain region only

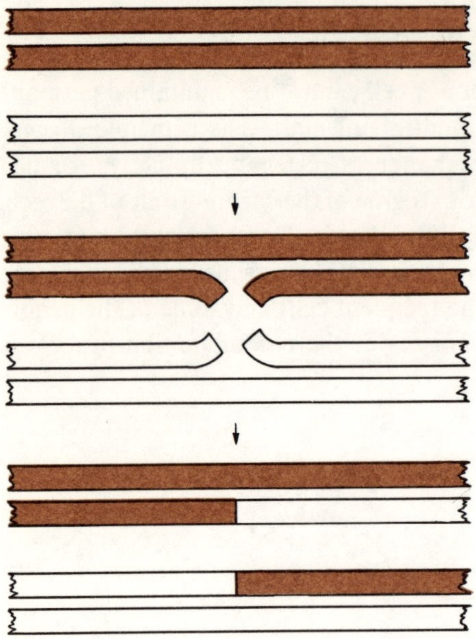

The first to reduplicate are nonsister chromatids.

Two nonsister chromatids are broken

and rejoined, giving rise to a reciprocal recombination.

Fig. 51. Crossing-over as the result of breaking and rejoining chromatids

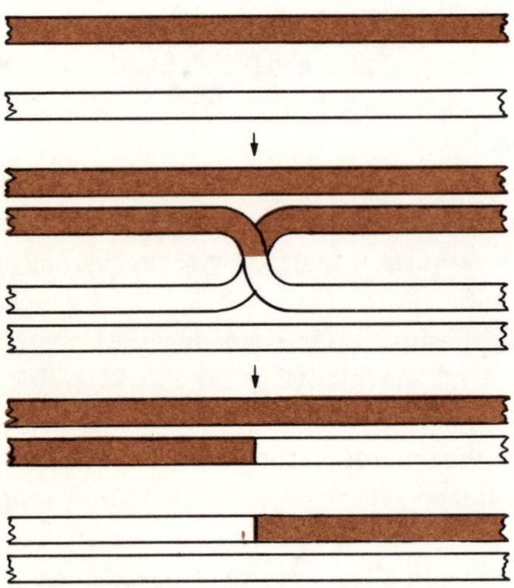

No reduplicated intact nonsister chromatids are formed.

Recombination is obligatorily coupled with chromatid reduplication. As new chromatids grow a shift occurs at a definite site and the nonsister chromatid begins to be copied.

Recombined chromatids are thus obtained without interruption of the chains.

Fig. 52. Crossing-over according to the copy choice theory

one chromatid is being read (Fig. 53). In regions indicated by arrows non-reciprocal exchange took place in both variants. Such non-reciprocal exchange results in a change of the segregation ratio* (Fig. 54).

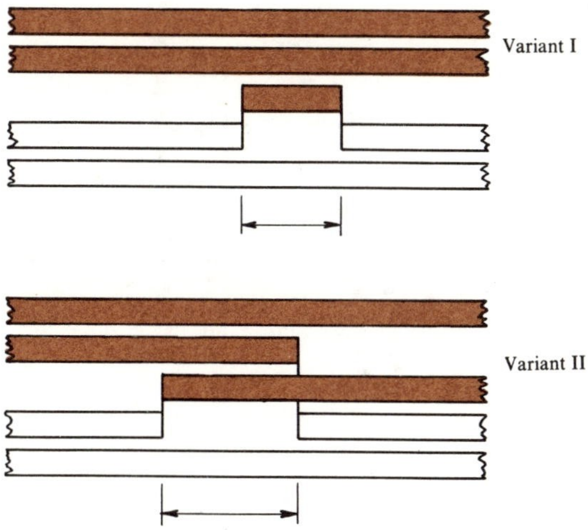

Fig. 53. First and second variant of nonreciprocal exchange of chromatids

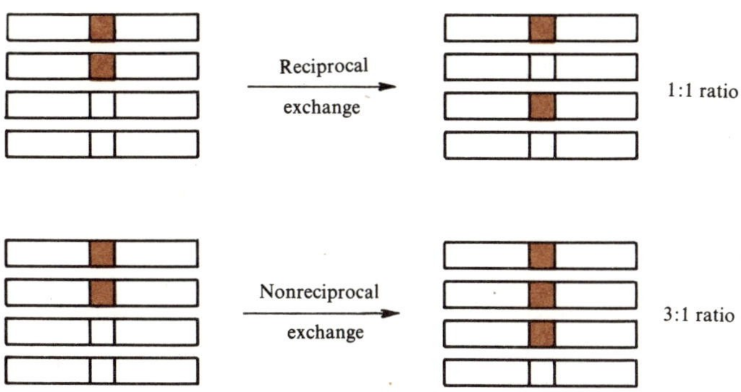

Fig. 54. Segregation ratio as a result of reciprocal and nonreciprocal exchange of chromatids

---

* In some ascomycetes the segregation ratio of a tetrad may be 5 : 3, probably owing to the presence of two subunits in the chromatids involved (Fig. 55).

Results obtained by isotopic studies of the recombination process of the λ phage suggest that the process involves breaking and rejoining double-helical DNA molecules rather than a copy-choice mechanism.

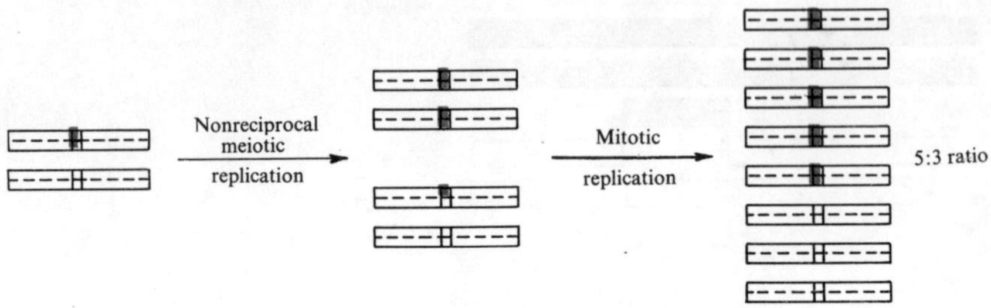

Fig. 55. Probable mechanism of formation of irregular segregation ratios in ascomycetes

It is clear that during some recombinations, such as those in bacteria, one of the mechanisms can be operative without giving rise to reciprocal recombinants. The recombination here involves the whole genome of the recipient cell but only a small part of the donor cell genome.

Gene Interactions

In the simplest case a given character of a diploid organism is determined by two alleles. Hence two different forms of the same gene interact during the phenotypic expression. An example of such allelic interaction is the phenotypic expression during incomplete dominance. The term gene interaction in a restricted sense of the word is reserved for relationships where the interacting genes (allelic pairs) are localized at different loci of the chromosomes (nonallelic interactions).

Gene interaction can occur at several levels: it can either proceed directly between the genes themselves or between the products of these genes or even at the level of metabolic interactions. It is one of the main tasks of studying gene interactions to define the nature of the interaction and to specify the level at which it takes place.

A typical example of gene interaction is the phenomenon of epistasis. During epistasis the effect of one gene, the epistatic one, is superior to that of other genes, the hypostatic ones. This gene subordination can manifest itself by a suppression of the phenotypic expression or even by an inhibition of the function of the hypostatic gene. However, the terms epistasis and hypostasis do not apply to interactions between paired alleles.

If a superior gene affects the expression of a certain allele by suppressing the expression of its mutant allele, it is known as a suppressor gene. Such a gene was mentioned in the section on suppressor mutation (p. 87).

Induction and repression of enzymes is based on gene interaction. The regulator gene localized outside the region of structural genes affects their function either negatively or positively.

A special case of gene interaction is the phenotypic expression of the so-called polygenes. A polygene is a gene which itself influences only slightly a certain phenotypic expression. However, this phenotypic expression may be quantitatively influenced by interaction of the polygene with a smaller or greater number of other equivalent genes. Such quantitative characters include some morphological dimensions and some physiological functions, crops of cultivated plants and yields of domestic animals, as well as the genetic control of the quantitative production capacities of microorganisms, such as yields of antibiotics.

Genetic studies of quantitative characters proceed from statistical characteristics and means of the values investigated, as Mendelian analyses of hereditary factors are not applicable here.

Extranuclear Inheritance

The decisive role of the nuclear genetic material has been amply discussed before. However, there exists a type of inheritance which is based on extranuclear genetic material. Extranuclear inheritance is thus governed by genetic factors which can exist and replicate outside the chromosomal structure. Mutagens affect such genetically active material similarly to the nuclear nucleic acids. Owing to the extranuclear localization of this type of inheritance, the distribution of the hereditary factors carried by extranuclear elements is irregular during both mitosis and meiosis*.

The origin, transfer and often the function of the carriers of extranuclear inheritance are still a matter of debate. The factors responsible are localized in mitochondria and chloroplasts but even outside these organelles, in the form of plasmids, in both eukaryotic and prokaryotic cells.

The name plasmid was introduced by Lederberg in 1952 as a genetic term for extranuclear genetic elements. Typical examples of plasmids are the F plasmid (p. 145), colicinogenic plasmids carrying the instruction for colicin synthesis, plasmids coding for resistance to antibiotics and toxinogenic plasmids. However, a plasmid is not just any molecule of extranuclear nucleic acid as it must be genetically homogeneous, have

---

* In higher organisms which reproduce sexually, the phenotypic expression of extranuclear inheritance is governed by the following criteria.

1. Segregation of extranuclearly controlled genes in the F_2 generation does not correspond to any type of Mendelian segregation, including gene interactions, sex-linked markers, etc.

2. In some instances matroclinity, i.e., maternal inheritance, is observed, the possible cause being that in the fertilization process not only the haploid nucleus of the egg but also its cytoplasm plays a role while the male gamete depends almost solely on its chromosomal genetic material. Matroclinity accounts for nonidentity of reciprocal crosses in the F_1 generation.

a constant size and be able to replicate outside the chromosome. There is no clear boundary between plasmids and phages or viruses but typical plasmids are clearly distinguishable from typical phage particles and it is possible that intermediate forms represent stages of evolution.

Many hereditary characteristics of higher organisms, as well as of microorganisms, are coded by extranuclear elements. Among microorganisms there is the interesting case of petite colonies of the yeast *Saccharomyces cerevisiae*. The frequency of occurrence of such petite colonies can be increased by applying a suitable mutagen.

The petite colonies arise from yeast cells which grow more slowly than the non-mutated strain because of lack of some mitochondrial cytochrome systems. The respiration-deficient strains do not segregate as normal "wild" cells do. Hybridization studies showed mitochondria to be the carriers of the petite character. However, it should be noted that the normal cytochrome equipment of this yeast is encoded in nuclear genes (for more details, see p. 83).

Analogous results were obtained with the ascomycete fungus *Neurospora crassa*. Its poky variants are phenotypically distinguished by a retarded growth of mycelium; the biochemical reason for this slow growth is a genetically based loss of some cytochromes.

Extranuclear inheritance can be caused by a large number of cytoplasmic factors of infectious nature. Thus, for instance, the kappa particles with a rickettsia-like structure are transmitted by contact. The kappa particles contain DNA and are subject to mutational changes. These infectious particles were found in *Paramecium aurelia,* specifically in strains which secrete paramecin into water. For individuals incapable of synthesizing paramecin the substance is toxic. Strains which form paramecin always contain a dominant nuclear gene K and cytoplasmic kappa particles. The transmission of this character is matroclinic.

A special group of plasmids are the so-called episomes, which differ from the rest by their ability to be integrated in chromosomes. They can replicate in two alternative states: either as an integral part of the bacterial chromosome or independently of it. In some species of *Enterobacteriaceae* a genetically coded conjugate resistance exists toward streptomycin, chloramphenicol, tetracyclines and sulphonamides. This resistance can be transferred from resistant cells to the sensitive ones on contact, through the genetically active factor of episomal type. The transfer of resistance need not be complete and can concern only some of the substances enumerated above. The plasmids of this type are called R (for resistance) factors and are assumed to have a circular structure capable of autoreplication. The R factors can exist either autonomously or integrated in the bacterial chromosome. Their molecular weight may range from $2 \cdot 10^6$ to $5 \cdot 10^8$.

Another episomal plasmid is the F (for fertility) factor and another still the genetic material of temperate phages, as will be discussed later.

The nature, transmission and function of episomal plasmids can be briefly described as follows:

1. Properties controlled by episomal plasmids are not vitally essential for a cell under normal conditions.

2. Spontaneous formation of episomal elements has never been observed. These factors can be acquired only through contact with cells containing the factor.

3. Episomes can exist in the host cell either in the cytoplasm as autonomous structures or they can be integrated in the bacterial chromosome. The autonomous form of the episome can change to the integrated form reversibly. An episome in the autonomous state can be transferred independently of the bacterial chromosome during conjugation. In the autonomous state it can sometimes be eliminated by the application of heavy metal salts, of acridine dyes or other interference with the host cell. In the integrated form the episome is relatively resistant to such elimination treatment.

4. Most episomes have the ability to mediate the transfer of bacterial genes from one cell into another.

Recombination Processes in Bacteria

Introduction

The explosive development of biochemical and molecular genetics was preceded by a long period of classical genetics, its principal biological object being higher organisms with a diploid number of chromosomes. Classical genetics was unable to reach the molecular level of genetic material even if the gene, considered as the fundamental autoreplicating unit of genetic analysis, already was accepted as a material particle.

In the 1940's, genetic investigations turned to microorganisms which offered unique possibilities of examining the inheritance of biochemical properties.

Avery and co-workers discovered in 1944 that foreign bacterial DNA can bring about changes in the hereditary properties of a cell. The technique of recombination with the aid of isolated DNA is called transformation. In 1946 a gene recombination was described that was based on conjugation of bacteria and in 1952 transduction was described as transmission of hereditary properties through bacteriophages. These three recombination mechanisms are often called parasexual mechanisms as they imitate to a certain degree the genetic processes taking place during sexual reproduction.

The various recombination processes in bacteria have the following common characteristics:

1. Recombinants are usually formed in a small fraction of the cells present; the frequency of recombinants can be substantially increased by special conditions.

2. The transfer of genes is unidirectional (from donor to recipient cells) and non-reciprocal.

3. The transiently formed zygotes are diploid for only a part of the transferred genetic material; they are called merozygotes.

The recombinant processes described could be observed with only a few bacterial species among them some actinomycetes.

The discovery of techniques permitting crossing of the bacterial genotype was particularly important for the theory of molecular genetics. Bacteria represent a particularly suitable biological object for such studies, owing to their uniqueness, the haploid state of their genome (which makes it possible even for recessive genes to be expressed phenotypically) and to the relatively simple structure of their chromosome. A substantial advantage of working with bacteria consists in the possibility of preparing large cell populations within a relatively short time interval. Using suitable selective cultivation conditions it is possible to detect even small numbers of mutant cells or recombinants within a numerous population.

Transformation

Transformation is the transfer of genetic information using DNA released from cells of the donor strain.

The discovery of transformation was preceded by the observation that killed cells can transmit certain characteristics to related bacterial strains. Griffith in 1928 was the first to study the problem with the aid of strains of *Diplococcus pneumoniae*. Encapsulated forms of this strain forming smooth colonies (S forms) are pathogenic. Pathogenicity is associated here with the presence of specific polysaccharides of the bacterial walls which act as antigens and can be distinguished serologically from strain to strain. The S forms give rise to colonies with a rough surface (R forms) which do not form the capsules and hence are not pathogenic. In his historic experiment, Griffith injected mice with a small amount of live cells of the R form along with a large number of heat-killed cells of the S form. The serological specificity of the bacterial capsules of the S strain was different from that of capsules of the S strain which had given rise to the R form used in the experiment. Mice injected with this mixture became infected and died. It was made sure that no living cells were present in the heat-inactivated culture of the S form. The mice died after infection caused by the virulent S form of pneumococci and the serological type of the capsular material of this S form was identical with the serological type of the capsules of the heat-killed culture. As it is known that spontaneous mutations do not result in changes of the antigenic character of the capsule, it was concluded that the ability to form the capsular antigen had been transferred to live R cells by a factor contained in the heat-inactivated culture.

Sixteen years after these experiments, the transformation-causing material was identified as DNA. The identification was based on the following evidence:

a) Exposure of the transformation material to deoxyribonuclease caused a loss of its transformation properties.

b) Ribonuclease or proteolytic enzymes had no effect on the transformation activity.

c) The penetration of transformation DNA labelled with ^{32}P into recipient cells was proportional to the number of transformants formed.

For a transformation to be successful the following must be observed:

1. Transformation DNA must possess a certain strand length. If the DNA is disintegrated (e.g., by ultrasonication or by the action of deoxyribonuclease) its short fragments are inactive in transformation. The minimum molecular weight of transformation DNA is usually 10^5. Single-stranded DNA has a smaller transformation effect than double-stranded DNA.

2. Transformation can be accomplished with cells existing in a certain physiological state called the competent state. Competence is checked by taking samples from a growing culture of the recipient strain and mixing them with an excess of transformation DNA. After several minutes, deoxyribonuclease is added and the number

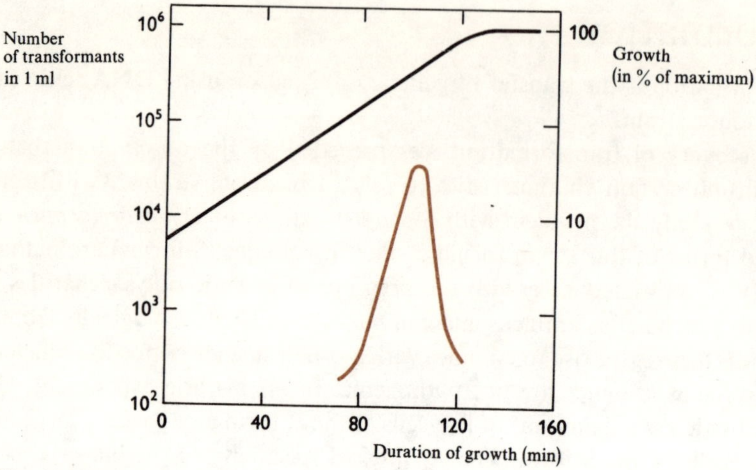

Fig. 56. Dependence of changes in cell competence on culture growth. Black, growth curve; coloured, number of transformants

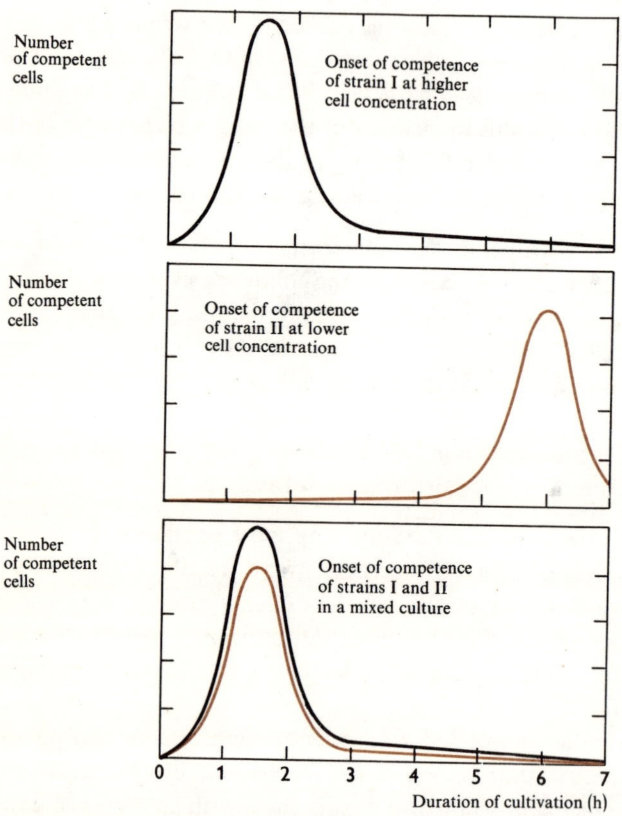

Fig. 57. Experimental proof of existence of the competence factor

of transformants is tested on a suitable medium. A typical result of such a test is shown in Fig. 56.

Of the many bacterial strains studied, only a few were found to be able to absorb DNA from the medium. The ability is evidently associated with the possibility of a rapid penetration of DNA through the several layers of the bacterial envelope. Competence appears rather suddenly during the growth of a culture, the ability rises abruptly to a maximum and then decreases again relatively rapidly. If a small inoculum is used for the cultivation competence will set in later than when a large inoculum is used. During mixed cultivation of two different strains with an unequal inoculum the appearance of competence of the low-inoculum strain is shifted exactly to the time of appearance of the high-inoculum strain (Fig. 57).

This phenomenon is accounted for by the production of a high-molecular-weight protein, the competence factor, the amount of which depends on cell concentration. On reaching a certain cell density the protein activator is rapidly transferred from cell to cell and competence appears as a kind of infection by the competence factor.

3. The number of transformants increases linearly with increasing concentration of transformation DNA. As a certain DNA concentration is reached (say, 200–300 molecules in *Bacillus subtilis*) the system becomes saturated and further increase of transformation DNA has no effect on the number of transformants (Fig. 58).

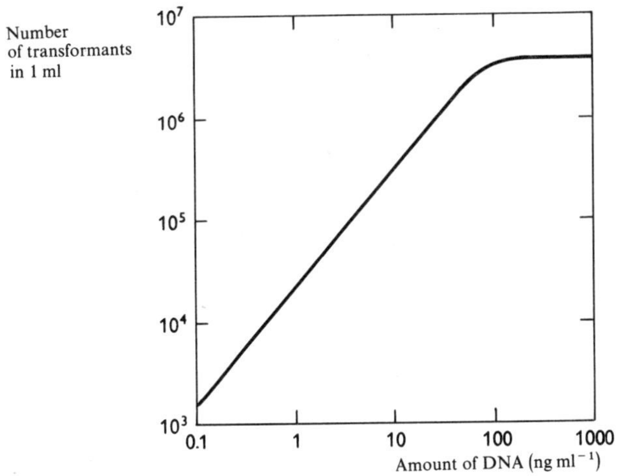

Fig. 58. Dependence between the number of transformants and the concentration of transformation DNA

4. Addition of a different type of DNA to the transformation DNA decreases the frequency of transformation. This effect is explained as being due to competition of two types of DNA for adsorption sites on the bacterial surface; for instance, a cell of *Diplococcus pneumoniae* contains about 50 adsorption sites for DNA.

Transfer of the ability to form cell capsules is not the only property that can be transmitted by transformation. The process can transfer the ability to synthesize some enzymes, the resistance to antibiotics, the ability to sporulate, the synthesis of protein antigens and the type of growth.

Transformation was successful with *Bacillus subtilis* and the genera *Neisseria, Streptococcus, Staphylococcus* and *Rhizobium.* In *Escherichia coli* transformation is only possible through an artificial generation of competence by calcium chloride under a defined temperature regime. This technique is applied in genetic engineering. Transformation is also known to take place in actinomycetes (see p. 164). Evidence of successful transformation in yeast is not yet satisfactory.

It is possible to bring about transformation between different related bacterial species but the frequency of such interspecies transformations is lower than within the species.

Transformation data revealed the linkage of some transferred characteristics, apparently due to the coexistence of genes for the two characteristics on the same fragment of isolated transformation DNA. Hence it is useful even for constructing a genetic map of bacteria.

In most transformations the donor cells can be used as recipient cells and vice versa. Hence it is known that in *Haemophilus* sensitivity and resistance to antibiotics can be transmitted both ways.

What happens to the DNA after it has been taken up by the bacterial cell? Transformation DNA apparently cannot be phenotypically expressed (i.e., it cannot be transcribed and subsequently translated). The first event is the formation of synapsis between endogenous DNA of the recipient cell and exogenous DNA of the donor cell (transformation DNA). At this stage the exogenous DNA is positioned on a certain complementary strand of the double-stranded chromosome of the recipient cell. The recombination process can then take place when a part of the transformation DNA is substituted for a part of the bacterial chromosome. Integration of a part of transformation DNA may cause a permanent change of the genotype of the transformed strain. Genetic information of the transformation DNA which has entered a cell can thus be expressed only after integration with the genome of the recipient cell. Such integration will occur 10 – 30 min after DNA penetration into the cell. According to some observations integration is accomplished only in bacteria formed by division of a cell that has received transformation DNA.

A special case of transformation is transfection where a phage or virus infection is transmitted to a competent recipient cell by a nucleic acid isolated from a virulent phage or virus.

Conjugation

Bacterial conjugation or fusion is a parasexual transfer of genetic material which requires direct contact between a live donor (male) and a live recipient (female) bacterial cell. During contact the cells are connected by a special tube through which passes a part of the donor cell chromosome (even a whole chromosome in some cases) into the recipient cell. For several generations the recipient cell may be diploid with respect to the transferred part of the genome as a merozygote. Recombinants are formed by a partial or a complete substitution of the donor chromosome for the homologous segment of the recipient chromosome in a process resembling crossing-over. Another haploid cell is formed, recombined in the corresponding segment of the genome.

Conjugation of bacteria thus begins by a contact of two cells of unequal mating type: the cells must have enough energy for building the connecting tube. A pair of conjugating cells of *Escherichia coli,* one with and the other without cilia, is shown in Fig. 59.

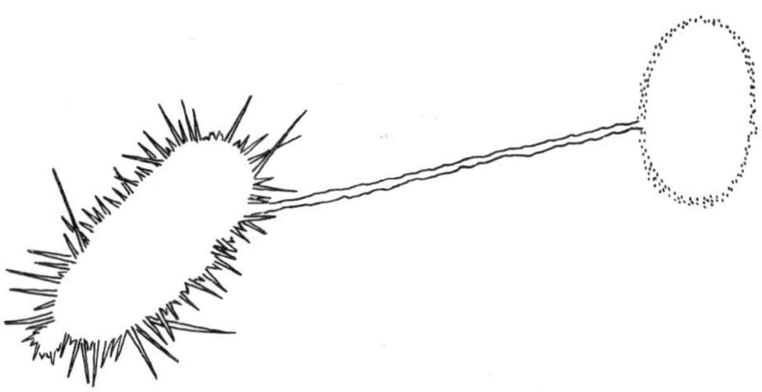

Fig. 59. Conjugating cells of *Escherichia coli*

The genetic material of the donor cell is transferred through the conjugation tube into the recipient cell. During conjugation only part of the donor chromosome is transferred as the relatively very long strand of chromosomal DNA is easily severed.

It was in 1946 that conjugation between two auxotrophic mutants of *Escherichia coli* K 12 was described. The auxotrophic strains were dependent on amino acids and vitamins as follows: I was $Met^-Bio^-Thr^+Leu^+B_1^+$; II was $Met^+Bio^+Thr^-Leu^-B_1^-$. There was thus no overlap in the dependences and the strains showed no spontaneous reversion to prototrophy. During joint cultivation of cells of both types in a complete liquid medium several prototrophs appeared, as shown schematically in Fig. 60.

Back mutation of the auxotrophs could be excluded as the cause of the phenomenon (with two factors of dependence such a reversion would have a frequency of 10^{-14}). Syntrophy was also eliminated (a situation when a colony contains cells of two auxotrophs mutually complementary in nutritional requirements) by plating the colonies on minimal and complete agar media. One possibility that was left was that recombination had taken place through transformation or transduction. Using a U-tube divided by a bacterial filter into two compartments where transformation or tranduction could take place on cultivating the two auxotrophs in the separate compartments this possibility was also excluded. The final interpretation of the observation was that genetic recombination took place through contact of cells of the two auxotrophs and that the genetic material was mutually exchanged (Fig. 61).

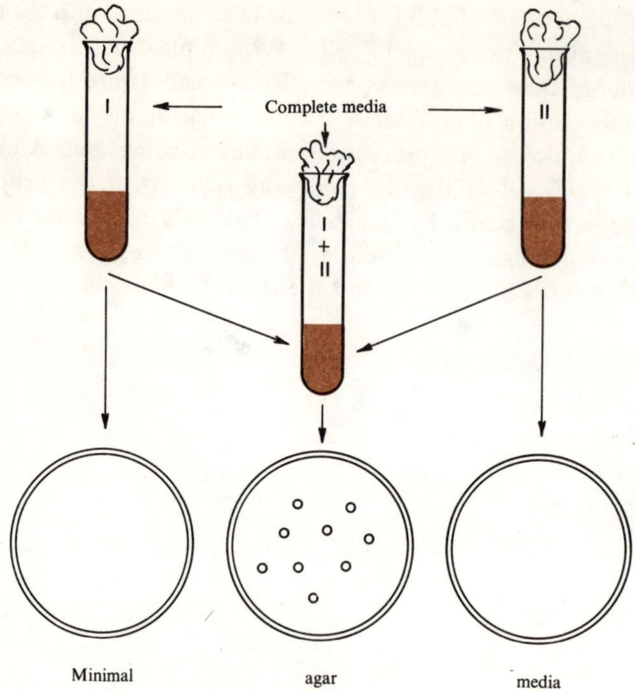

Fig. 60. Outline of a conjugation experiment

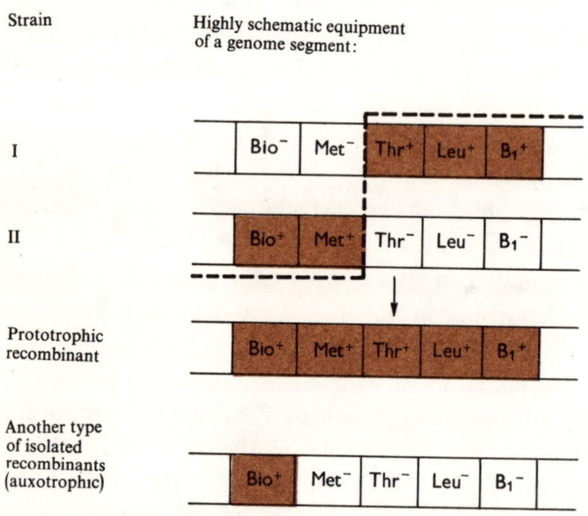

Fig. 61. Examples of recombinations in auxotrophic mutants

The mechanism of conjugation was clarified in 1952. *Escherichia coli* K 12 was found to possess a certain type of heterothallism, reflecting the existence of donor and recipient cells in a culture. Cells with the donor properties contain the F (fertility) factor and behave as male cells. The F factor was already mentioned as an autonomous episomal plasmid (p. 135). It is not a firmly fixed structure and it can be eliminated (unless integrated in the chromosome) by salts of cobalt and nickel, by acriflavin or by simple aeration of cells in the stationary phase. Cells containing the F factor are designated F^+ and those without it (female, recipient cells) F^-. The M_r of the F factor is $4.5 \cdot 10^7$. Its presence in the cell also affects its surface properties.

During crossing F^+ with F^- strains a small number of recombinants is obtained but strains exist which bring about a high frequency of recombinations (Hfr strains):

$F^- \times F^-$ no recombinants
$F^+ \times F^+$ very few recombinants
$F^+ \times F^-$ 10^{-6} recombinants
Hfr $\times$ F^- 10^{-2} recombinants

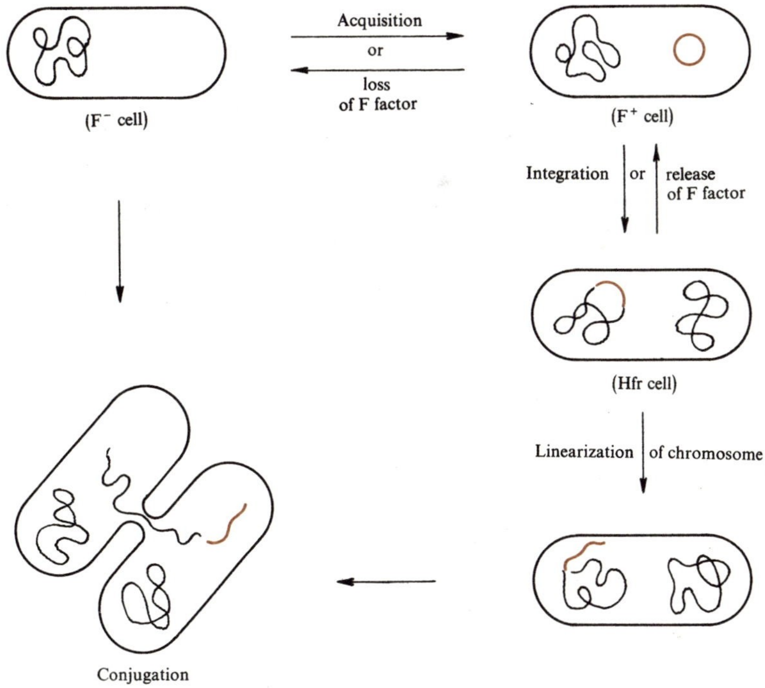

Fig. 62. Function of the F factor in the formation of Hfr strains, in chromosome linearization and in conjugation

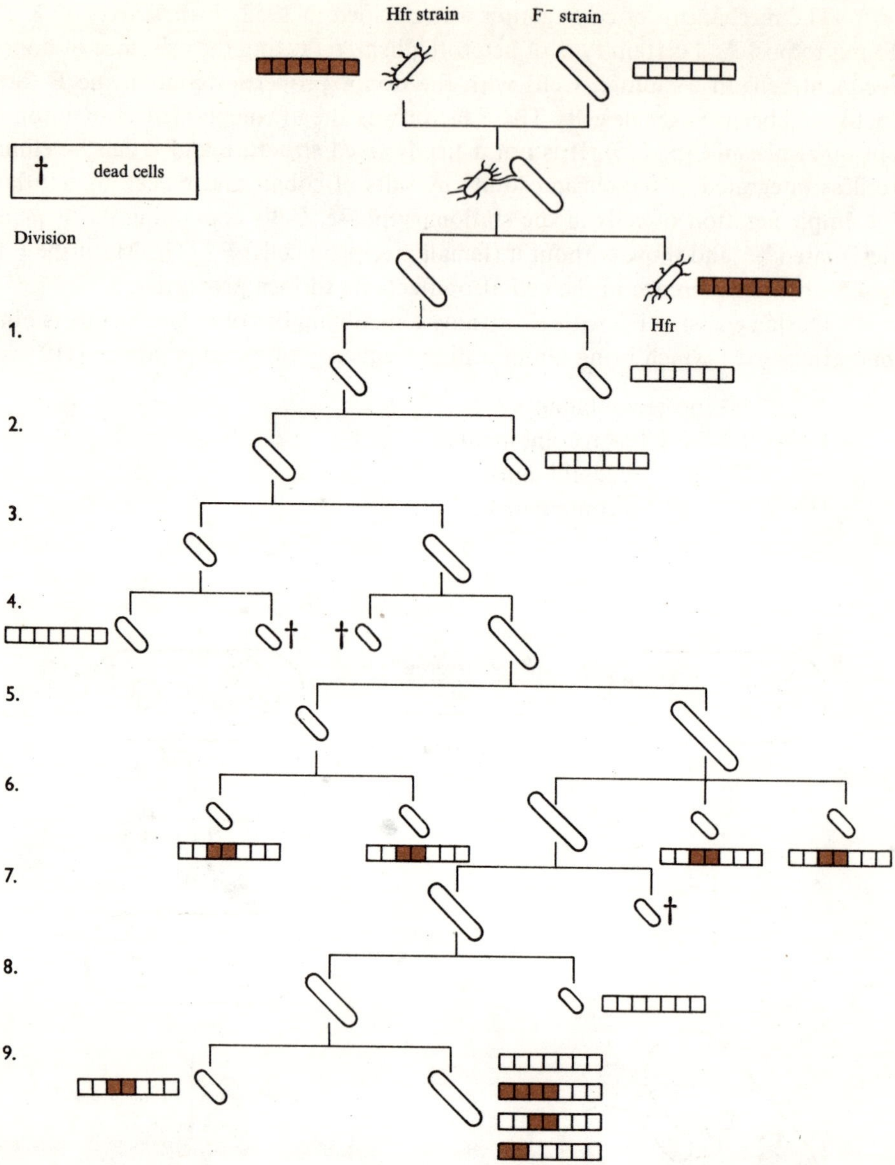

Fig. 63. "Family tree" of *Escherichia coli* cells after conjugation.
The Hfr cell is short and motile, the F^- cell is larger and not ciliated. The Hfr strain is defined as (O)$B_1^+Leu^+Az^sT5^sLac^+Gal^+Sm^s$ and the F^- strain as (O)$B_1^-Leu^-Az^rT5^rLac^-Gal^-Sm^r$. The symbols are explained in the legend to Table IX. Each gene is represented by a small square

The F factor can be joined to the bacterial chromosome. This joining can cause an opening (linearization) of the circular chromosome at the site of joining with the F factor. The F factor then governs the migration of the linearized chromosome into the recipient cell: first to enter the recipient cell is the DNA terminus without the F factor attached which is called the origin (O). The Hfr strains contain the F factor integrated in their chromosome.

DNA migrating from the donor to the recipient cell is single-stranded but is converted to the normal double-stranded form in the recipient cell. Single-stranded DNA is formed by replication of the circular bacterial chromosome by the rolling circle mechanism (p. 42). This replication apparently drives the single-stranded DNA directly into the recipient cell. The mutual relationships of the processes are depicted schematically in Fig. 62.

Detailed information on processes through which the cell progeny passes after conjugation was obtained by following the "genealogy" of a single pair of cells of *Escherichia coli* Hfr × F^-. Using a micromanipulator a pair of conjugating cells was isolated and the fused cells were separated and followed for several generations. The Hfr partner remained viable after conjugation in most instances and the F^- partner was viable in all instances after conjugation. The family tree of cells after conjugation is shown in Fig. 63. The genotype of the individual cells was determined from the properties of colonies growing out of these cells. Recombinant cells appeared only after six divisions. One out of two cells after the ninth division gave rise to colonies containing cells with four genotypes: one was pure F^- and the other three were recombinants with mixed properties of Hfr and F^- strains.

When using cells containing the F factor, all the genetic characteristics of the donor cells are transferred with the same probability. This may be accounted for by the fact that the F factor can be attached to the circular chromosome at any random site. However, if Hfr strains are used there is a strikingly high frequency of certain characters in the recombinants. The preference of these characteristics depends on the Hfr strain used, apparently because in the Hfr strains the F factor is integrated in the bacterial chromosome at a definite and unique locus which may be different for different strains (scheme in Fig. 64). The frequency of characters will decrease in strains I–IV in the following sequence:

I: **f** > **e** > **d** > **c** > **b** > **a** > **h** > **g**
II: **d** > **c** > **b** > **a** > **h** > **g** > **f** > **e**
III: **b** > **a** > **h** > **g** > **f** > **e** > **d** > **c**
IV: **g** > **f** > **e** > **d** > **c** > **b** > **a** > **h**

The transfer of the linearized chromosome into the recipient cell was analysed in detail by interrupting conjugation by stirring rapidly in a homogenizer. DNA was thus shown to enter into the recipient cells in a linear sequence. The DNA strand, which is extremely long is comparison with the cell size, enters at a rate of 10^5 deoxyribonucleotides per minute.

Conjugation was successfully used for preparing a genetic map of the chromosome of *Escherichia coli* K 12. For this purpose, several Hfr strains of *Escherichia coli* K 12 were prepared with differences in the position of the origin which is the first to enter the recipient cell. With the aid of these strains it was possible to map the position of the genes lying next to O, and by combining the results from several Hfr strains to construct the map. The map provides evidence for the circular (closed) structure of the bacterial chromosome.

An example of the genetic map obtained with the aid of Hfr strains is shown in Fig. 65, which depicts the chromosome of *Escherichia coli* K 12. The outer circle shows the sequence of genes without respect for the absolute distances between them. Genes with imprecise localization with respect to neighbouring genes are shown in parentheses.

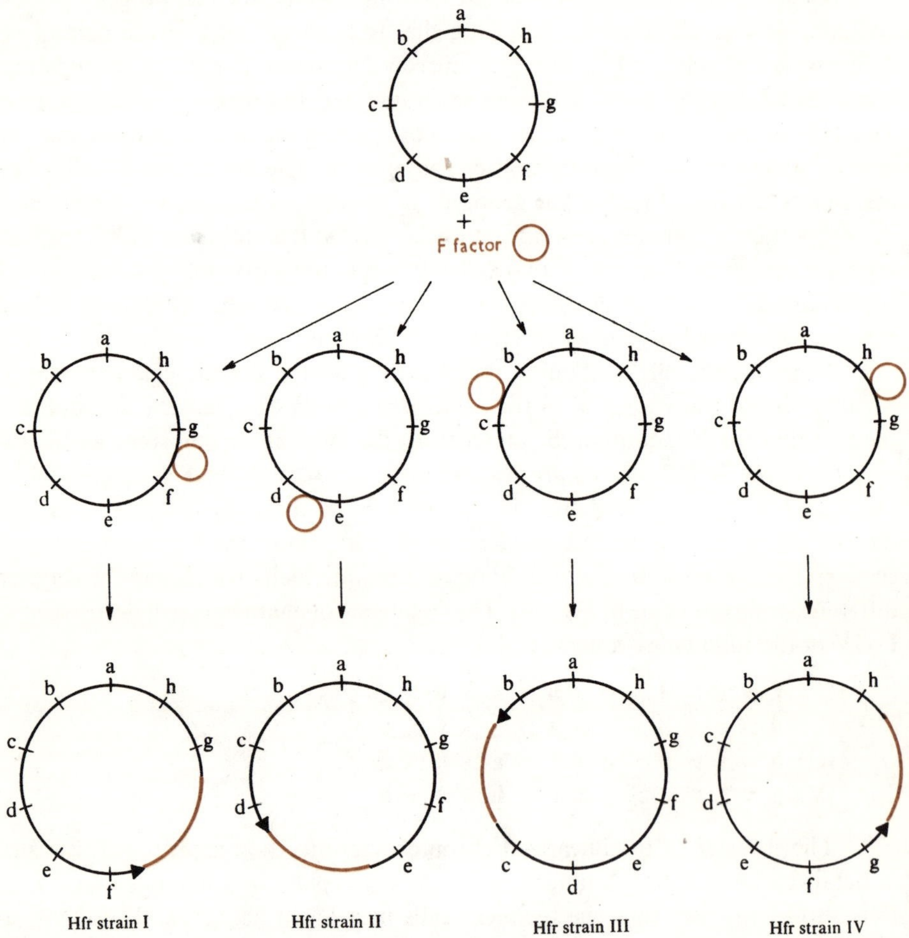

Fig. 64. Mechanism of formation of different types of Hfr strains

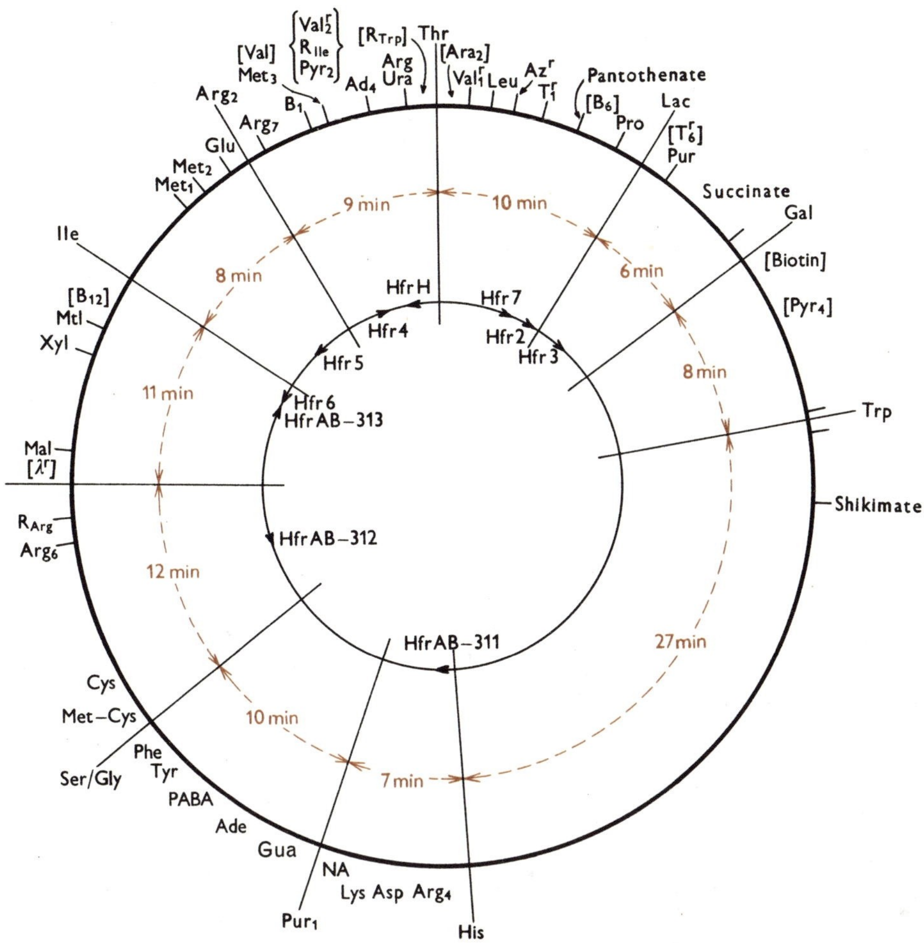

Fig. 65. Simplified map of the *Escherichia coli* K 12 chromosome.
The symbols are the same as in the legend to Table IX. Further gene symbols are the following: Synthesis of proline (Pro), pyrimidines (Pyr), arginine (Arg), lysine (Lys), nicotinamide (NA), guanine (Gua), adenine (Ade), *p*-aminobenzoic acid (PABA), tyrosine (Tyr), phenylalanine (Phe), cysteine (Cys), vitamin B_{12} (B_{12}) and valine (Val); fermentation of arabinose (Ara); dependence on succinic acid (Suc), aspartic acid (Asp) and glutamic acid (Glu); resistance to valine (Val^r) and phages (T^r, λ^r); repression by arginine (R_{Arg}), isoleucine (R_{Ile}) and tryptophan (R_{Trp})

The middle dashed circle shows the time intervals in which the characters shown on the outer circle were transferred by conjugation. The inner circle shows the origins (arrows) of the individual Hfr strains used.

It is now known that the linearized chromosome enters the recipient cell starting with the origin. The sequence of characteristics experimentally established for individual Hfr strains is shown schematically in Table IX. This genetic map also contains genes localized in other studies, not mentioned in the table.

Table IX. Sequence of genes in different Hfr strains of *Escherichia coli*

					Type of Hfr strain					
H	1	2	3	4	5	6	7	AB 311	AB 312	AB 313
O	O	O	O	O	O	O	O	O	O	O
Thr	Leu	Pro	Pur$_2$	B$_1$	Met	Ile	T1^r	His	Sm^r	Mtl
Leu	Thr	T1^r	Lac	Met	B$_1$	Met	Az^r	Trp	Mal	Xyl
Az^r	B$_1$	Az^z	Pro	Ile	Thr	B$_1$	Leu	Gal	Xyl	Mal
T1^r	Met	Leu	T1^r	Mtl	Leu	Thr	Thr	Pur$_2$	Mtl	Sm^r
Pro	Ile	Thr	Az^r	Xyl	Az^r	Leu	B$_1$	Lac	Ile	Ser/Gly
Lac	Mtl	B$_1$	Leu	Mal	T1^r	Az^r	Met	Pro	Met	His
Pur$_2$	Xyl	Met	Thr	Sm^r	Pro	T1^r	Ile	T1^r	B$_1$	Trp
Gal	Mal	Ile	B$_1$	Ser/Gly	Lac	Pro	Mtl	Az^r	Thr	Gal
Trp	Sm^r	Mtl	Met	His	Pur$_2$	Lac	Xyl	Leu	Leu	Pur$_2$
His	Ser/Gly	Xyl	Ile	Trp	Gal	Pur$_2$	Mal	Thr	Az^r	Lac
Ser/Gly	His	Mal	Mtl	Gal	Trp	Gal	Sm^r	B$_1$	T1^r	Pro
Sm^r	Trp	Sm^r	Xyl	Pur$_2$	His	Trp	Ser/Gly	Met	Pro	T1^r
Mal	Gal	Ser/Gly	Mal	Lac	Ser/Gly	His	His	Ile	Lac	Az^r
Xyl	Pur$_2$	His	Sm^r	Pro	Sm^r	Ser/Gly	Trp	Mtl	Pur$_2$	Leu
Mtl	Lac	Trp	Ser/Gly	T1^r	Mal	Sm^r	Gal	Xyl	Gal	Thr
Ile	Pro	Gal	His	Az^r	Xyl	Mal	Pur$_2$	Mal	Trp	B$_1$
Met	T1^r	Pur$_2$	Trp	Leu	Mtl	Xyl	Lac	Sm^r	His	Met
B$_1$	Az^r	Lac	Gal	Thr	Ile	Mtl	Pro	Ser/Gly	Ser/Gly	Ile

Explanation of symbols: origin (O), synthesis of threonine (Thr), leucine (Leu), proline (Pro), purines (Pur$_2$), tryptophan (Trp), histidine (His), serine and glycine (Ser/Gly), methionine (Met), isoleucine (Ile), and thiamine (B$_1$); fermentation of lactose (Lac), galactose (Gal), maltose (Mal), xylose (Xyl) and mannitol (Mtl); resistance to sodium azide (Az^r), streptomycin (Sm^r) and phage T1 (T1^r).

Conjugation is not a strictly species-specific process but interspecies transfer of the F factor proceeds at a lower frequency. An example of such conjugation is the transfer of the F factor from *Escherichia coli* into a strain of *Salmonella.* Salmonellae of the F$^+$ type that are thus formed, then conjugate with F$^-$ types of salmonellae at the usual frequency of 10^{-6} per cell.

Sex-duction or F-duction is based on the incorporation of adjacent bacterial genes into the F factor and their subsequent transfer through conjugation. This is possible because during the separation of the F factor from the bacterial chromosome a certain part of the chromosome may be taken with it and will then migrate together with the F factor into the recipient cell. The altered (enriched) F factor, which is designated F′, then replicates autonomously with the transferred part of the chromosome. This represents an analogy to the transfer of a bacterial chromosome with the aid of temperate phages as discussed below.

Transduction

Transduction is a recombination process in which part of the bacterial chromosome is transferred by a phage.

Phages belong to the category of viruses which are known to be infective particles incapable of their own reproduction, parasitizing at the genetic level on other living systems. They are divided according to the type of their nucleic acid, into RNA viruses and DNA viruses, or according to the type of host, into vertebrate and invertebrate viruses, vertebrate viruses, plant viruses, insect viruses and bacterial viruses (bacteriophages, phages).

A modern classification divides phages into eight groups:

1. T-Even phages (T2, T4, T6) containing a double-stranded DNA and $M_r = 1.2 \cdot 10^8$; their tail is contractile.

2. λ-Phages with double-stranded linear DNA and $M_r = 3.1 \cdot 10^7$; their tail is long and non-contractile.

3. Phages of group PM 2 with circular double-stranded DNA and $M_r = 5 \cdot 10^6$; they have no tail.

4. Phages of group ΦX174 with circular single-stranded DNA and $M_r = 1.7 \cdot 10^6$; they form small spherical virions.

5. Phages of group fd with circular single-stranded DNA and $M_r = 1.7 \cdot 10^6$; they form long filamentous virions.

6. Phages of group f2 with linear single-stranded RNA and $M_r = 1.2 \cdot 10^6$; they appear as small virions.

7. Phage T7 and related types with linear double-stranded DNA and $M_r = 2.5 \cdot 10^7$ to $2.7 \cdot 10^7$; they contain a short tail.

8. Phages of group Φ6 with linear double-stranded RNA in three copies with $M_r = 1.3 \cdot 10^7$; their virions have a lipid capsule.

The morphology of the phage particle is demonstrated on phage T4, which infects cells of *Escherichia coli*. Fig. 66 shows that the most prominent feature of the phage is its head, composed of proteins in which the genetic material is located. The head is joined to a hollow tail covered in a sheath and terminating in a hexagonal plate. This plate then carries tail fibres and short spikes; 60% of the mass of this phage is represented by protein and 40% by nucleic acid. The enzymic equipment of phages is very poor, reflecting their parasitic form of life. Some enzymes are present which assist in the perforation of the bacterial cell wall (Fig. 66) and enzymes of the contractile apparatus of the tail sheath (phosphatases). The phage proteins possess antigenic properties – an animal organism generates specific serum antigens after injection of the phage particles.

DNA of the T4 phage differs from other known DNAs by containing 5-hydroxymethyldeoxycytidylic acid instead of deoxycytidylic acid; glucose is bound to the 5-hydroxymethylcytosine residue in the molecule of phage DNA.

Infection of a bacterial cell with a phage has several stages. The first involves the attachment of the virion to the cell surface, a process assisted by the tail fibres which are first coiled around the tail end and are unwound before phage adsorption to the cell. The bacteriophage is fixed to the cell surface with the aid of the spikes. Receptors permitting the adsorption of the phage are localized in the bacterial cell wall; bacterial protoplasts possessing no cell wall are not attacked by the phage.

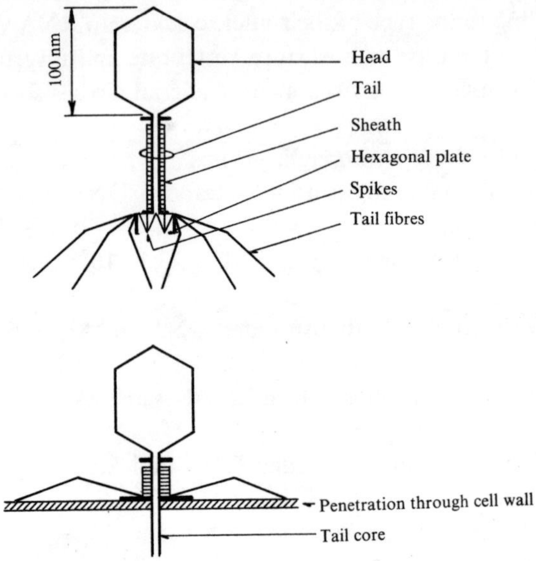

Fig. 66. Schematic morphology of a phage and technique of bacterial cell wall perforation

In a following phase the bacterial cell wall is perforated by lysozyme contained in the hexagonal plate of the phage. Disruption of the peptidoglycan component of the cell wall creates an opening into which the core of the phage tail is inserted. This is assisted by a contraction of the sheath. DNA from the phage head is then injected through the tail core. Other parts of the phage particle remain outside the infected cell.

As soon as the phage DNA penetrates into the bacterial cell the bacterial chromosome is fragmented, synthesis of bacterial proteins ceases and proteins controlled by the phage genes begin to be synthesized by the cell machinery.

The first proteins thus synthesized are enzymes required for the replication of phage DNA; their synthesis sets in as early as 5 min after cell infection. After a further 3 min a second set of phage genes comes into play. These code for the structural proteins of the phage and this starts the morphogenesis of the phage, which is now known in some detail from the use of T4 mutants. Fig. 67 shows it to proceed in three independent branches, one involving the formation of head particles, the second that of the tail and the third that of the tail fibres. The phage is then composed by an attachment of the tail to the head and finally by joining the tail fibres. A discrete mechanism is responsible for this sequence which cannot be altered.

The first complete phage particle appears within 13 min after cell infection. In 25 min from the beginning of infection a cell will contain as many as 200 phage particles. Under the action of phage lysozyme the cell wall is perforated from the inside and phages are released into the medium to spread invasively toward other bacterial cells.

According to their virulence, phages can be divided into the virulent group (those that bring about complete cell lysis) and the temperate group. If a suspension of virulent phages is mixed with a bacterial culture in a suitable ratio the turbid bacterial culture will completely clear as a consequence of complete cell lysis. The suspension thus formed is known as phage lysate. The number of phage particles is usually deter-

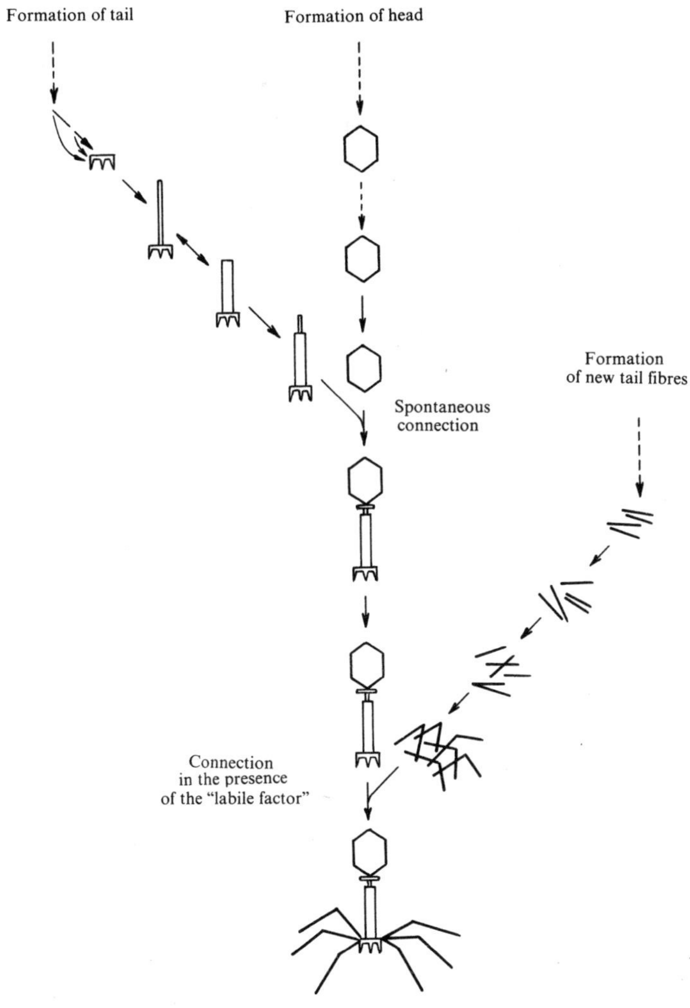

Fig. 67. Formation of a phage particle

mined by the so-called plaque method on agar media. A concentrated suspension of sensitive bacterial cells is mixed with a small number of phage particles and poured on an agar plate. After a certain period of incubation the continuous and opaque bacterial film on the agar will begin to show small translucent areas known as plaques. Each plaque is the result of lytic action of phages stemming from a single phage in the original mixture*. From the dilution of the phage lysate used for the test one can calculate the number of phages in it. It should be observed that the appearance of the plaques is characteristic of the phage used.

The other type of phage is the temperate phage. Such phages can bring about lysis of cells just as the virulent ones do. However, they can pass to a stage of prophage in which case the phage nucleic acid is attached to the bacterial chromosome and replicates together with it. No cell lysis takes place and the situation is described as lysogeny. Integration of the phage DNA in the bacterial genome which causes lysogeny prevents further synthesis of phage proteins. This regulatory mechanism is based on a principle which is analogous to that encountered during the control of synthesis of induced and repressed enzymes in bacteria. The lysogenic state is maintained by the production of repressors which are bound to the operator gene and block the function of structural genes which, in the uninhibited state, control the synthesis of proteins required for building up the phage particles. A lysogenic bacterial culture is immune to infection by homologous phages which can inject their DNA into lysogenic cells, but this DNA is inactivated by the repressor produced by the prophage and is eliminated during subsequent cell divisions. Lysogenic bacteria inherit the ability to produce the phage of the same genotype as the one that has caused the lysogeny. Certain factors (such as UV light and some chemical agents) can be applied to induce the phage to reproduce. It is separated from the bacterial chromosome, begins to replicate to form other phage particles and lysis eventually takes place.

A temperate phage can be used during transduction as the vector which transfers small fragments of the bacterial chromosomes from the donor to the recipient strain. The transduction process is illustrated in Fig. 68.

The auxotrophic strain placed in the left arm of the U-tube (Fig. 68) contains a temperate phage which may cause lysis in some of the cells, be released into the medium and penetrate through the bacterial filter to cells of the other auxotroph placed in the right arm of the U-tube. These cells now begin to be lysed and the phage, enriched with DNA fragments of this strain, passes back to the left arm of the tube. Some of the cells present there undergo transduction so they receive, together with the phage genome, some fragments of the bacterial chromosome. The strain in the left arm acts as a recipient and that in the right arm as a donor. It is now well established that the vector in the transfer of genetic information described here is really a bacteriophage.

---

* This phage particle caused a number of consecutive lytic cycles and a single plaque may contain as many as 10^7 phages.

For a transduction to be successful, a phage must be used such that some of the cells survive the infection and become lysogenic or prevent phage multiplication by some other mechanism. The donor strain cells must be partly lysable by the phage used.

After receiving phage DNA with attached fragments of the bacterial chromosome, the lysogenic cells turn into merozygotes where genetic recombination is possible. The recombinants formed are called transductants.

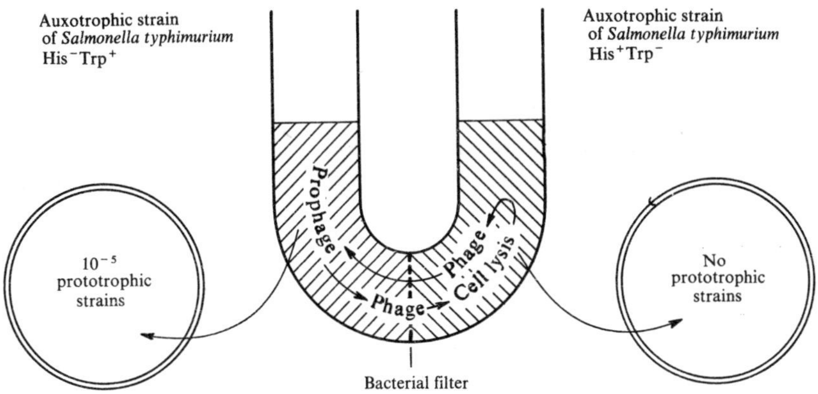

Fig. 68. Experimental arrangement for following transduction

The following types of transduction are known:

1. Nonspecific transduction. The transducing phage can transfer genes from any region of the donor cell chromosome into the recipient cell culture. The absolute transfer frequencies are low (about $10^{-5}-10^{-4}$ transductant per phage particle). During the transfer of genetic information only tightly linked genes are transferred (the so-called co-transduction). Nonspecific transduction has been described with *Salmonella typhimurium, Escherichia coli, Pseudomonas aeruginosa, Proteus mirabilis, Bacillus subtilis* and *Staphylococcus aureus*.

2. Specific transduction. The phage can transfer genes only from a certain region of the donor chromosome. This type of transduction was observed in *Escherichia coli* K 12, where the temperate λ phage transfers only genes of the galactose region into recipient cells.

3. Abortive transduction. During nonspecific transduction it may happen that no synapsis is formed between the recipient chromosome and the part of the donor chromosome. In such cases no permanent recombinants are formed. The introduced chromosomal fragment can remain in the cytoplasm of the recipient cell and influence its phenotype. As the cell divides, the DNA fragment passes into only one of the daughter cells and is thus gradually diluted out of the culture.

Nonspecific and abortive transduction have been used for a fine genetic analysis of the bacterial chromosome.

Genetic Engineering

Introduction

Genetic engineering consists of the use of techniques which make it possible to transfer genes from one organism to another by overcoming the interspecies barrier.

Controlled manipulations with genes are based on the use of a set of enzymes, called restriction endonucleases, which act on DNA as a fine chemical scalpel and split it specifically. Very often this cleavage gives rise to DNA fragments with single-stranded sticky ends. Using another enzyme, one can join the sticky ends through covalent bonds. Still other enzymes can be used for cutting out and processing the required gene fragments from the donor genome.

An important tool of genetic engineering has been found in plasmids (p. 135) or temperate phages (p. 154) which are used for cloning the genes. After an enzymically catalysed incorporation of a certain gene sequence into the plasmid or phage structure the modified plasmid or phage acts as a vector for entering the bacterial cell in which it multiplies to a large number of copies including the incorporated sequence. When such transformed bacteria are plated on an agar medium the desired clone can be isolated by a monocolonial selection of bacteria. Such gene cloning may also be called molecular cloning.

From the point of view of practical application it appears now to be most attractive to incorporate animal genes into bacterial cells and to obtain thus by "fermentation" some highly prized compounds, such as insulin and some other hormones, interferon, growth factors and immunoglobulins.

Genetic engineering provided techniques for transferring genes coding for ribosomal RNA in the African frog *Xenopus laevis* into plasmids multiplying in *Escherichia coli.* Likewise, genes coding for histones in sea urchin eggs were transferred into *Escherichia coli.* Rabbit and mouse genes for globin and the mouse gene sequences for immunoglobulin were incorporated into plasmid DNA and plasmids were prepared containing genes for rat insulin. Recently, commercial production of human insulin by transformed bacteria has started. Synthetic gene for human neurohormone somatostatin (14 amino acids) introduced into a bacterial cell now also serves for the production of this hormone.

Enzymes Used in Genetic Engineering

The main group in this category are the restriction endonucleases or restrictases of microbial origin. Especially useful are those endonucleases which split only those segments of DNA which exhibit a two-fold symmetry of the bases, e.g.,

5′...GAATTC...3′
3′...CTTAAG...5′

The two strands of DNA are seen to contain the same sequence but running in opposite directions. Such inverse sequences are common in DNA and are called (incor-

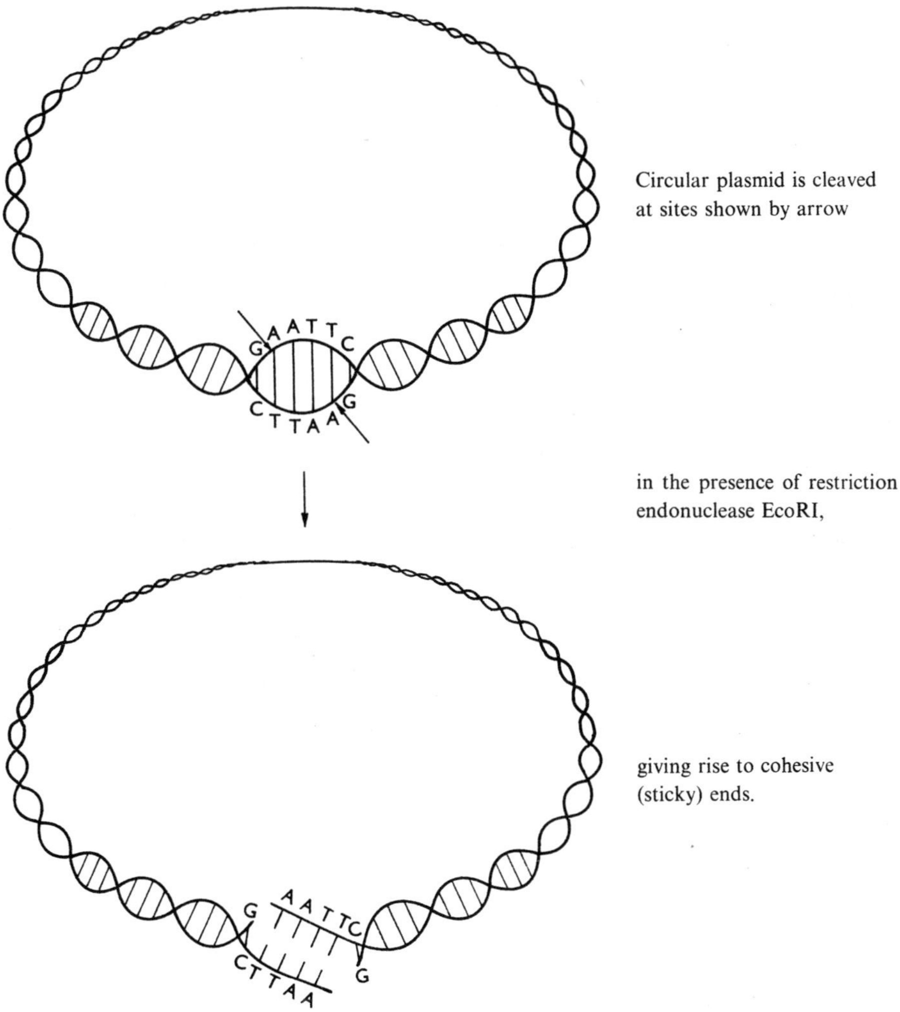

Fig. 69. Schematic representation of linearization of plasmid DNA with the aid of restriction endonuclease

rectly) palindromes. Cleavage of the above palindrome by the bacterial restriction endonuclease EcoRI is illustrated in Fig. 69.

Restriction endonucleases are important not only for genetic engineering but also for determining the nucleotide sequences (the so-called sequencing) of DNA as they permit to isolate short segments of DNA which are amenable to analysis. It is now possible to do more or less routine sequencing of polynucleotides with up to 300 nucleotide residues.

Very important enzyme is the reverse transcriptase (p. 58), which can transcribe the information contained in mRNA into DNA, and DNA polymerase, catalysing the synthesis of double-stranded DNA from single DNA strands (p. 38).

A covalent bond between DNA strands is formed by DNA ligase while undesirable single-stranded ends can be removed with S1 nuclease prepared from *Aspergillus oryzae*. Another enzyme of importance is terminal transferase which catalyses the prolongation of the 3′-termini of double-stranded fragments of DNA with shorter single-stranded segments.

Transfer of the Globin Gene into a Bacterial Cell

The example of introducing the animal gene for globin into bacterial cells will serve for explaining, in a simplified form, one of the more complicated techniques of genetic engineering. The three principal phases of the procedure are shown in Fig. 70A, B and C.

In the first phase the circular DNA of the pCRI plasmid from *Escherichia coli* is exposed to the action of the restriction enzyme EcoRI which splits it at a single site, giving rise to an open plasmid DNA with cohesive (sticky) ends. This DNA is then exposed to DNA polymerase in the presence of dATP and dTTP, which turns it into a fully double-stranded molecule. With the aid of terminal transferase and in the presence of dCTP as substrate, the 3′-termini of the DNA are both extended by oligodeoxycytidylic chains (Fig. 70A).

In the second phase (Fig. 70B), the eukaryotic gene for globin must be prepared. The mRNA for globin is isolated from rabbit reticulocytes where this particular mRNA species is a predominating component of the total. The isolated mRNA then serves as a template for reverse transcriptase (RNA-dependent DNA polymerase) which generates single-stranded DNA from low-molecular-weight precursors. After removal of RNA this single-stranded DNA is exposed to *Escherichia coli* DNA polymerase (again in the presence of all the precursors) to form double-stranded DNA. Action of S1 nuclease removes the hairpin structure from it and thus creates the globin gene. Terminal transferase with dGTP as substrate then extends both 3′-termini of this DNA by oligodeoxyguanylate chains.

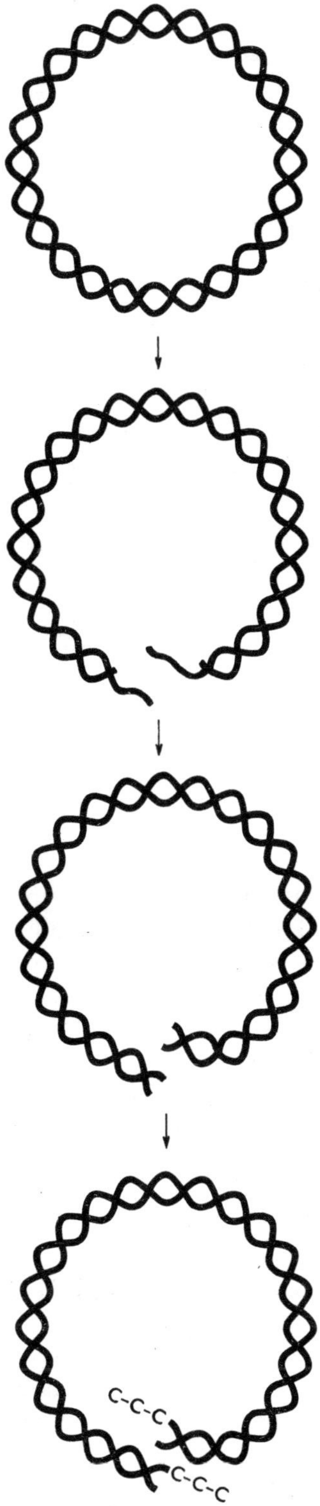

A plasmid

is cleaved by restriction endonuclease specifically

to form a linear structure with single-stranded ends.

Using DNA polymerase

the single-stranded ends are complemented
to double-stranded ones

and in the presence of terminal transferase and dCTP

the 3′-terminal ends of both strands are extended by chains of oligodeoxycytidylate (cohesive ends).

Fig. 70A. Preparation of plasmid DNA with cohesive ends

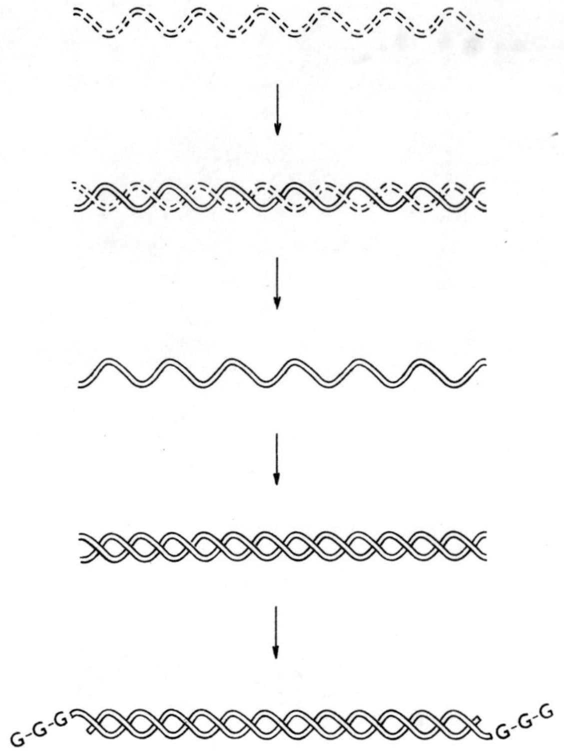

mRNA from rabbit reticulocyte coding for the synthesis of globin serves as template

for reverse transcriptase *in vitro* to form

a complementary strand of DNA.

Denaturation will remove RNA and the remaining

single-stranded DNA

serves as template for DNA polymerase to form

a double-stranded molecule of DNA – the gene for globin.

Using terminal transferase and dGTP

the 3′-terminal ends of both strands are extended by oligodeoxyguanylate chains (cohesive ends).

Fig. 70B. Preparation of globin DNA with cohesive ends

In the third phase (Fig. 70C), the globin gene is introduced into the plasmid structure. The two types of DNA prepared in phases I and II are mixed and hybridization of the terminal sequences containing oligodeoxycytidylate (in the plasmid DNA) with those containing oligodeoxyguanylate (in the globin DNA) takes place. Polynucleotide ligase (usually isolated from bacteria infected with the T4 phage) generates covalent bonds at the ends of heterologous DNAs and a combined circular plasmid is formed. This plasmid then invades (by the transformation process) a bacterial cell, such as that of *Escherichia coli*.

This was the technique that was used for creating bacterial strains containing α- or β-globin gene sequences.

Synthesis of Insulin by Transformed Bacteria

Insulin mRNA was isolated by a special technique from Langerhans islets of rat pancreas and transcribed into complementary DNA using reverse transcriptase from the avian myeloblastoma virus. The mRNA was then removed from the DNA complex

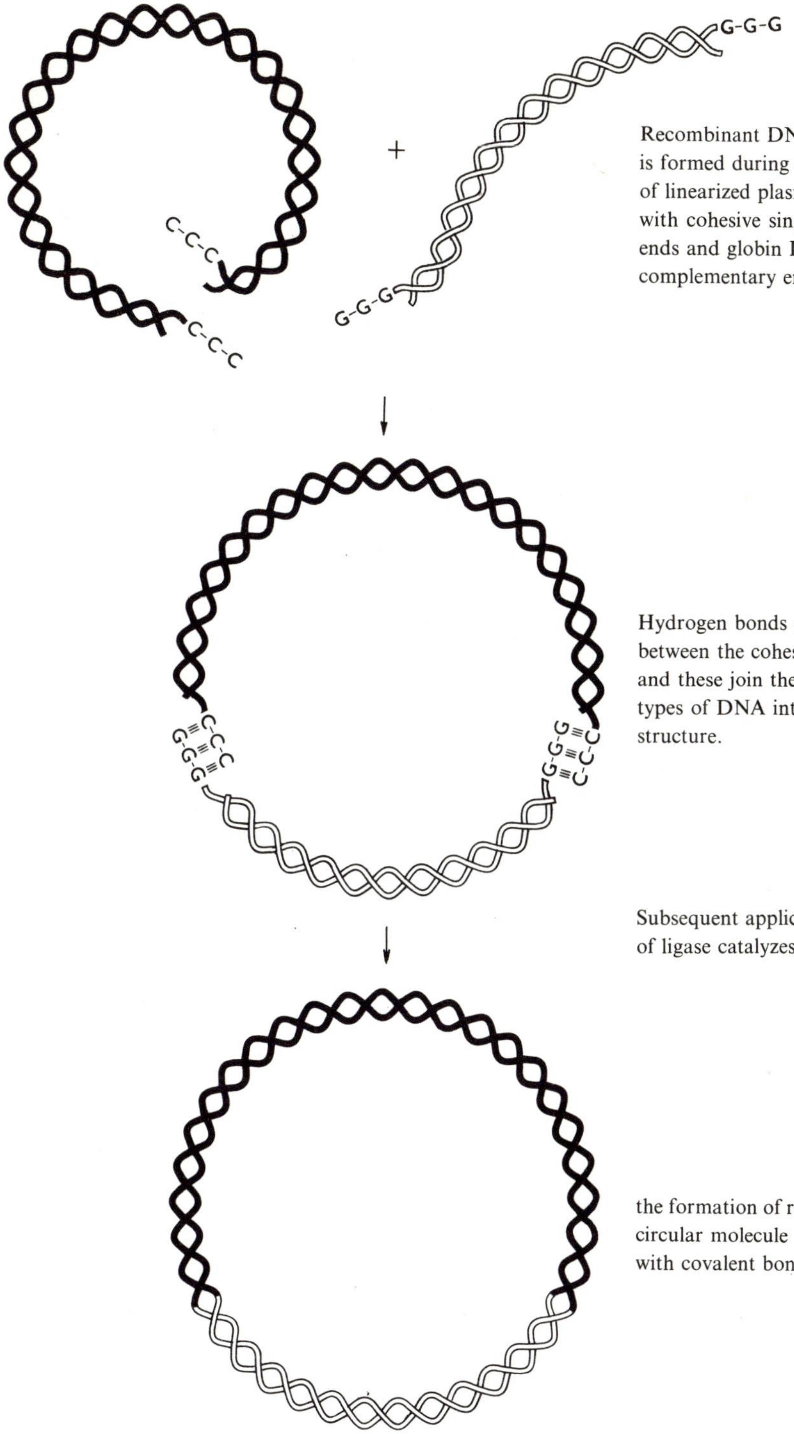

Recombinant DNA is formed during contact of linearized plasmid DNA with cohesive single-stranded ends and globin DNA with complementary ends.

Hydrogen bonds are formed between the cohesive ends and these join the two types of DNA into a circular structure.

Subsequent application of ligase catalyzes the formation of recombinant circular molecule with covalent bonds.

Fig. 70C. Preparation of recombined DNA

in an alkaline medium and double-stranded DNA was obtained again through the activity of reverse transcriptase. This DNA has the characteristic long hairpin structure. This fraction was then exposed to S1 nuclease and restriction endonuclease. The fragments formed were cloned, starting with an enzyme-catalysed attachment of a restriction ligand prepared synthetically (the decanucleotide 5′CCAAGCTTGG3′, which contains a sequence cleavable by the HindIII restrictase). The fraction exposed to this restrictase was then purified by polyacrylamide gel electrophoresis.

The insulin gene thus obtained was incorporated into the pMB9 plasmid of $M_r = 3.5 . 10^6$, which causes resistance to tetracycline in *Escherichia coli.* This plasmid can also be split by restrictase HindIII at a single site, giving rise to cohesive ends.

Plasmids containing the insulin gene were used to transform cells of *Escherichia coli.* The transformed cells were then selected on a tetracycline-containing medium.

Modern methods of synthesis of human insulin make use of "synthetic genes" for the A and B chains of insulin, i.e. polydeoxyribonucleotides prepared by chemical synthesis. The beginning of each of these genes carries the coding triplet for methionine, the end is terminated by a stop codon. The modified synthetic genes were introduced separately into the gene for β-galactosidase carried by a suitable plasmid vector. The recombined plasmid was then introduced into cells of *Escherichia coli* where a polypeptide is formed, composed of a part of β-galactosidase and of an insulin chain (A or B), joined together by a methionine residue. Application of cyanogen bromide destroys the linking methionine (insulin itself does not contain methionine). The released chains of insulin A and B are then isolated and joined by disulfide bonds, giving rise to the final molecule of human insulin.

Outline of the Genetics of Actinomycetes

Although some actinomycetes have long been used as producers of antibiotics, anticoccidial substances and other biologically active compounds, the present state of understanding of their genetics is unsatisfactory, the apparent reason being that actinomycetes are much more difficult to manipulate than simple bacteria, such as *Escherichia coli.*

Streptomycetes are the group of actinomycetes that have found special application in microbial production. On agar media they form colonies resembling those of fungi but cytologically and genetically they are undoubtedly prokaryotic and hence kin to bacteria. Long, fine and branched hyphae of the vegetative mycelium form compact colonies on solid media and these give rise to aerial mycelium with large multigenomic cells, the hyphae then forming chains of spores. Cytology revealed that streptomycete hyphae can fuse. In young hyphae the genetic material occurs in the form of long rods which fill in most of the space between septa. In older hyphae the genetic material appears as shorter and shorter rods until it becomes spherical and is stored in the spores, one small sphere per spore. There is no nuclear envelope and in many other respects the genetic material of streptomycetes resembles that of bacteria, including the fact that it is haploid.

The complex organization of streptomycetes complicates the evaluation of recombination experiments and practically excludes the use of some recombination techniques.

In spite of this, many recombination studies have been performed even here. If two auxotrophic strains of actinomycetes with different nutritional requirements are grown in a mixed culture on complete agar, prototrophic recombinants are obtained at a frequency of $10^{-5}-10^{-4}$. The presence of recombinants is examined in spores obtained after the mixed cultivation and subsequent sporulation of the parent strains, by plating the spores on selective agar media.

Recombination Processes in *Streptomyces coelicolor*

When plasmids had been discovered in various strains of streptomycetes, their genetics began to look more promising. Fundamental in this respect were studies on *Streptomyces coelicolor*, which can contain a plasmid either as an autonomous unit (the strain is designated $SCP1^{+}$) or in an integrated form (SCP1); a plasmid-free strain is then described as $SCP1^{-}$.

The recombination studies done with *Streptomyces coelicolor* yielded the following frequencies of recombinants (in %):

$SCP1^-$	× $SCP1^-$	0.001
$SCP1^+$	× $SCP1^+$	0.01
$SCP1^+$	× $SCP1^-$	0.01
SCP1	× SCP1	1
SCP1	× $SCP1^+$	10
SCP1	× $SCP1^-$	100

The extremely high frequency found in the last case reflects the fact that all the spores obtained contained in their circular chromosome the incorporated plasmid genome. The relatively high frequency of recombinants obtained on crossing $SCP1^-$ with $SCP1^+$ is due to the presence of another plasmid (SCP2); without this other plasmid the recombinant frequency would be several orders of magnitude lower.

It follows from the above that there is a certain analogy between the behaviour of the *Streptomyces coelicolor* plasmid and of the F factor in *Escherichia coli*, although differences exist. During crossing of $SCP1^-$ with $SCP1^-$ the recombination process is not polarized so that the genetic material can be transferred both ways, without any overal distinction between a donor and a recipient cell. The transfer of the chromosome is equally nonpolarized and every part of it has the same probability of being transferred. In analogy, $SCP1^+$ crosses with $SCP1^+$ and SCP1 with SCP1. On the other hand, crossing of SCP1 with $SCP1^+$ or with $SCP1^-$ is polarized: in both instances the SCP1 strain acts as a donor (analogy with the Hfr strain of *Escherichia coli*). It is essential that the plasmid is also transferred into the recipient cell. During such recombinations of actinomycetes a special nature of the gradient of frequencies of the transferred genes was observed, the gradient decreasing in both directions from the integrated plasmid. It thus appears that the plasmid is transferred every time and that it carries different lengths of the actinomycete chromosome with it. The highest probability of transfer is then found with genes directly adjacent to the integrated plasmid. These findings led eventually to a relatively successful mapping of the chromosome of *Streptomyces coelicolor*.

Modern Techniques of Recombination of Actinomycetes

Thermoactinomycetes attain a transformation frequency of as much as 10^{-3}. As with the transformation process in *Escherichia coli*, the competence property appears at a certain phase of growth. In actinomycetes, transduction using a temperate actinophage has been performed.

With mesophilic actinomycetes conjugation has been observed during direct contact of cell mycelium used for recombination. The importance of the plasmid as a sex factor in *Streptomyces coelicolor* was mentioned above.

In other actinomycetes plasmids play a role during the production of antibiotics (chloramphenicol, tetracyclines, macrolides and others) and of melanin. However, it is not clear whether in these instances the plasmids can also assume the role of the sex factor.

The existence of plasmids in actinomycetes is gaining in importance as it opens up possibilities of manipulating the production capacities of these microorganisms by genetic engineering techniques. It should be noted, however, that the presence of plasmids has been demonstrated mostly in indirect ways – through the sensitivity of some plasmid-coded properties to treatment with ethidium bromide, acridine orange, etc.

There is a new recombination technique now available, based on the fusion of cell protoplasts, the fusion being greatly stimulated by polyethylene glycol and short electrical pulses applied in an electrophoretic field. It is to be expected that interspecies crossing of actinomycetes will become possible by the application of the fusion techniques.

Actinomycetes are known for the occurrence of the so-called heteroclones, which represent a mixture of recombinant and parent genotypes and readily segregate to heterogeneous populations of haploid strains. Heteroclones are a valuable model for studying dominance and complementation as they permit a diploid state of the chromosome to be attained at least partly, in the otherwise strictly haploid actinomycetes.

Heterokaryons have also been described in actinomycetes that will grow on a selective medium, but they are known to segregate readily into the two parental genotypes. The basis of heterokaryosis in actinomycetes is not well understood. It is not clear why two parental genomes can persist together in the same cytoplasm without intense mutual recombination.

Principles of the Genetics of Fungi

Introduction

The large taxonomic group of fungi, including the yeasts and moulds, are eukaryotic organisms, the nucleus of which resembles in its complicated structure the nuclei of higher organisms; likewise, the structure of fungal chromosomes is much more complicated than the bacterial chromosome equivalent and corresponds to that of multicellular organisms. In diploid cells of fungi the chromosomes occur in homologous pairs. Nuclear divisions (mitosis and meiosis) take place in fungal cells just as they do in animal or plant cells, although other than optical evidence had to be provided for this statement. The possibility of sexual reproduction makes it possible in most fungal species to carry out hybridizations with useful genetic consequences which are analogous to the consequences of hybridization in plants and animals.

In bacterial genetics the macromolecular approach to the genetic material was sufficient for explaining practically all the phenomena; in contrast, with fungi the complex rules of heredity of eukaryotic cells must be taken into consideration, as discussed in detail in the theoretical part of this book. However, to comprehend the genetics of fungi, what is also required is the morphology of vegetative and reproductive structures, ways of sexual reproduction and types of cell cycles. In some fungi (such as yeasts and moulds) there is a complicated organization and movement of genetic material and the advantages of the genetic model encountered in bacteria are still preserved. Among the advantages one should mention first the unicellularity of most representatives or at least of their spores, the haploid nature of their nuclei and the relatively simple cultivation procedure which permits the use of rapid selection techniques.

Obviously this chapter cannot provide a full and complex account of fungal inheritance, but an attempt will be made to present the principal problems of fungal genetics and to introduce some of the experimental approaches, in particular for those representatives of this large group of organisms which are interesting from the viewpoint of food chemistry and biotechnology.

Life Cycles of Fungi

Two types of reproduction exist in fungi, sexual and asexual (vegetative). With representatives of ascomycetes, basidiomycetes and phycomycetes, both types occur in parallel. Haploid and diploid nuclear phases alternate, just as with other sexually reproducing higher organisms. The artificially constructed group *Fungi imperfecti* includes those fungal species in which only vegetative reproduction has ever been observed. In addition to many yeasts, the group also includes some species of *Aspergillus* and *Penicillium*.

The vegetative stage of fungi (but not of yeasts) is characterized by an elongated, branched, mainly dikaryotic to polykaryotic cells, capable of forming filaments or hyphae. A tuft of such hyphae is the mycelium. Asexual reproduction is effected either by fragments of hyphae or by asexual spores, such as conidia or oidia. Normal cells of hyphae and conidia mainly contain haploid nuclei but often more than one.

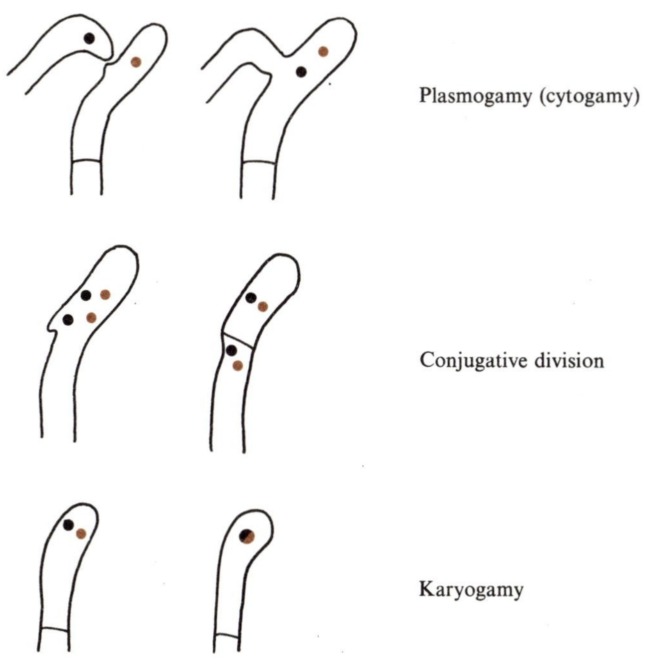

Fig. 71. Initial stages of ascus formation

The sexual process of fungi begins by fusion of two sexually compatible cell elements (spores, vegetative cells or differentiated sexual organs), each of which contains one or more haploid nuclei. Such fusion is described as plasmogamy or cytogamy. The nuclei of such fusing elements regain their individuality and associate in one or more pairs, each of the pairs representing a dikaryon. Dikaryotic cells undergo mitotic division, giving rise to daughter cells of the same dikaryotic equipment (conjugative division). The period of the dikaryotic phase may differ from species to species and may even be absent altogether. The sexual process itself is then manifested by a fusion of the paired nuclei in the dikaryotic cell (karyogamy), which sets the stage for the diploid phase of the cell's life cycle (Fig. 71).

The diploid nucleus formed by karyogamy then undergoes meiosis. The haploid products formed are the basis of future sexual spores (ascospores, basidiospores). The situation is best illustrated by the example of ascus formation in the terminal ascogenic hypha of *Neurospora* (Fig. 72).

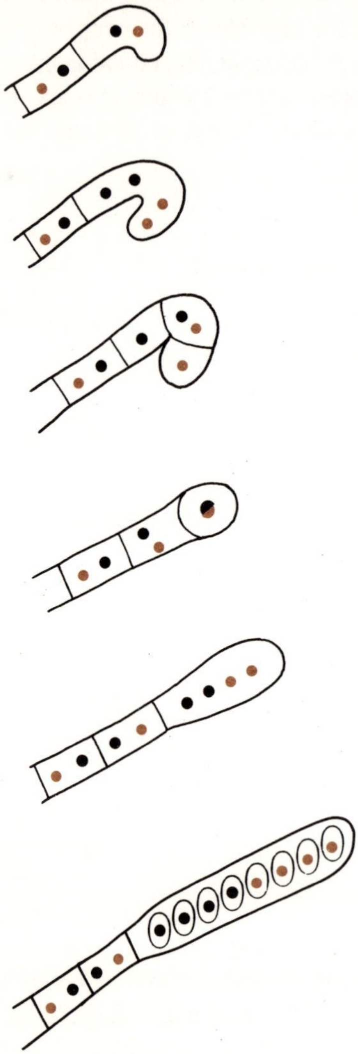

An ascogenic hypha forms a dikaryotic cell at its tip.

The two nuclei divide by mitosis and a hook-shaped tetrakaryotic cell is formed.

In a later stage the terminal dikaryotic cell is separated and forms the basis of a future ascus.

The haploid nuclei of this cell then fuse into a single diploid nucleus (karyogamy). At the same time, the two monokaryotic cells fuse into a dikaryotic one.

The diploid nucleus undergoes meiosis and simultaneously the ascus is extended.

Meiotic division forms four nuclei which divide again by mitosis so that eight haploid nuclei of future ascospores are created.

Fig. 72. Development of ascus in *Neurospora*

For most of their life fungi exist as haploid organisms (although many yeasts are generally diploid) in which the nuclei fuse (karyogamy) at the end of the vegetative stage. The resulting diploid undergoes meiotic division to haploid sexual spores. The life cycle of the sexually reproducing fungi thus proceeds from a fully haploid state to a fully diploid state. The duration of these states differs widely from group to group, and this fact gave rise to a classification of fungal cell cycles into seven basic types. Following Raper and his types A – G, this is shown in Fig. 73. Table X shows the characteristics of the A – G cycles and the main representatives of fungi are shown in which the given life cycle occurs.

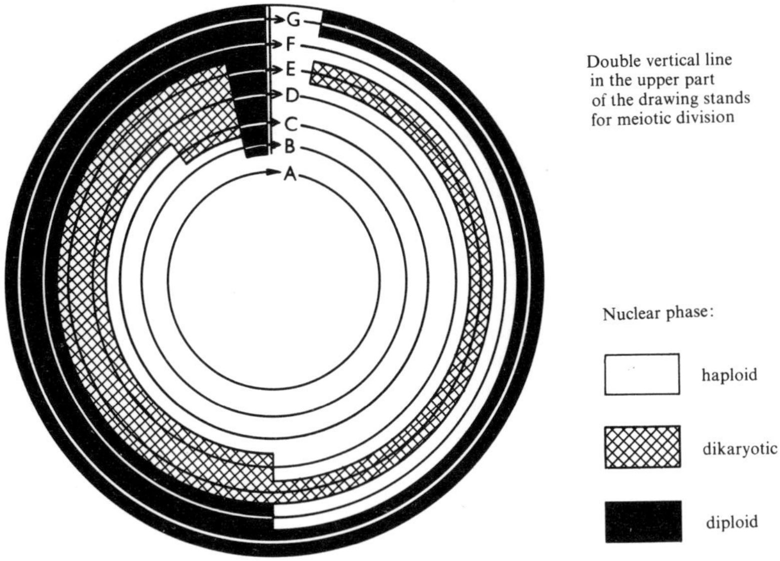

Fig. 73. Diagram of life cycles of fungi

Among the *Fungi imperfecti*, where no sexual reproduction is possible, a true hybridization is not feasible. Their asexual cycle gives rise only to clones. With many species of the group, however, genetic recombination can be achieved by hybridization of the parasexual type which requires the formation of a heterokaryon, i.e., a strain with dikaryotic cells, each of the nuclei carrying a distinct genome.

Life Cycles of Yeasts

The life cycles of yeasts, especially those of interest in industry, will now be treated in some detail. Yeasts are often divided into haploid and diploid depending on whether the longest part of their life cycle is characterized by haploid or diploid nuclei. However, careful analysis reveals that some yeast representatives exhibit a brief dikaryotic phase. Like other fungi, yeasts can be classified in seven basic types of life cycles.

The type A cycle (asexual) is exemplified by the so-called haploid yeasts, such as *Torulopsis, Candida, Rhodotorula* and other *Cryptococcaceae*. These yeasts re-

Table X. Principal life cycles of fungi according to Raper (see also Fig. 73)

Cycle	Characteristic cycle phase	Characteristic features	Chief representatives
A	asexual	No sexuality and no change of karyotic phase. Exclusively vegetative reproduction.	*Fungi imperfecti.* Some species of *Aspergillus* and *Penicillium*, producing only conidia from spores. Also haploid yeasts.
B	haploid	Reproduction both sexual and asexual. Vegetative phase fully haploid. Zygote is formed immediately before meiosis.	Most members of *Phycomycetes*, lower genera of *Ascomycetes* (some yeasts).
C	haploid, limited duration of dikaryotic stage	Reproduction both sexual and asexual. Predominant part of vegetative phase haploid. Diploid stage is preceded by brief dikaryotic phase.	Higher ascomycetes, such as genus *Neurospora.*
D	haploid-dikaryotic	Reproduction both sexual and asexual. Vegetative phase is half haploid, half dikaryotic. Diploidization takes place immediately before meiosis.	Most fungi belonging to *Basidiomycetes.*
E	dikaryotic	Reproduction both sexual and asexual. Dikaryotic vegetative stage which represents the main karyotic phase begins by fusion of products of meiosis (ascospores or basidiospores). Both haploid and diploid phase are strongly reduced.	Many smuts.
F	haploid-diploid	Reproduction both sexual and asexual. Haploid vegetative stage passes directly into the diploid stage. No dikaryotic phase.	Algae and higher plants, exceptionally fungi (*Allomyces*).
G	diploid	Reproduction both sexual and asexual. Haploid state restricted to immediate products of meiosis. Vegetative stage is fully diploid. No dikaryotic phase.	Most yeasts, *Myxomycetes,* some *Blastocladiales.*

produce solely asexually. The diploid phase is unknown and many strains of this family are now considered as degenerate forms of diploid sporogenic yeasts.

The type B life cycle (haploid) is exemplified by genera *Coccidiascus, Debaryomyces, Endomyces, Endomycopsis, Lipomyces, Nematospora, Pichia, Schizosaccharomyces, Schwanniomyces, Wingea* and *Zygosaccharomyces.* These yeasts reproduce either sexually or asexually. They spend most of their life as haploid cells, the diploid state being limited to the zygote. The zygote here plays the role of the ascus in which haploid spores are formed.

Life cycles C, D and E are characterized by a dikaryotic phase of different length.

The dikaryotic phase is usually very short (such as in *Nadsonia*) but relatively stable dikaryotic phases are known from genera *Sporidiobolus* and *Rhodosporidium*.

Yeasts with a type F life cycle (haploid-diploid), which has only two stages, a haplophase and a diplophase of about equal duration, are represented by *Hansenula wingei*.

The life cycle of type G (diploid) is represented mainly by *Saccharomyces* species. Most of the vegetative phase of these cells is diploid, the haploid one being relatively short. Many strains of *Saccharomyces cerevisiae* have a tendency toward spontaneous diploidization. In *Saccharomyces chevalieri* the spontaneous diploidization is control-

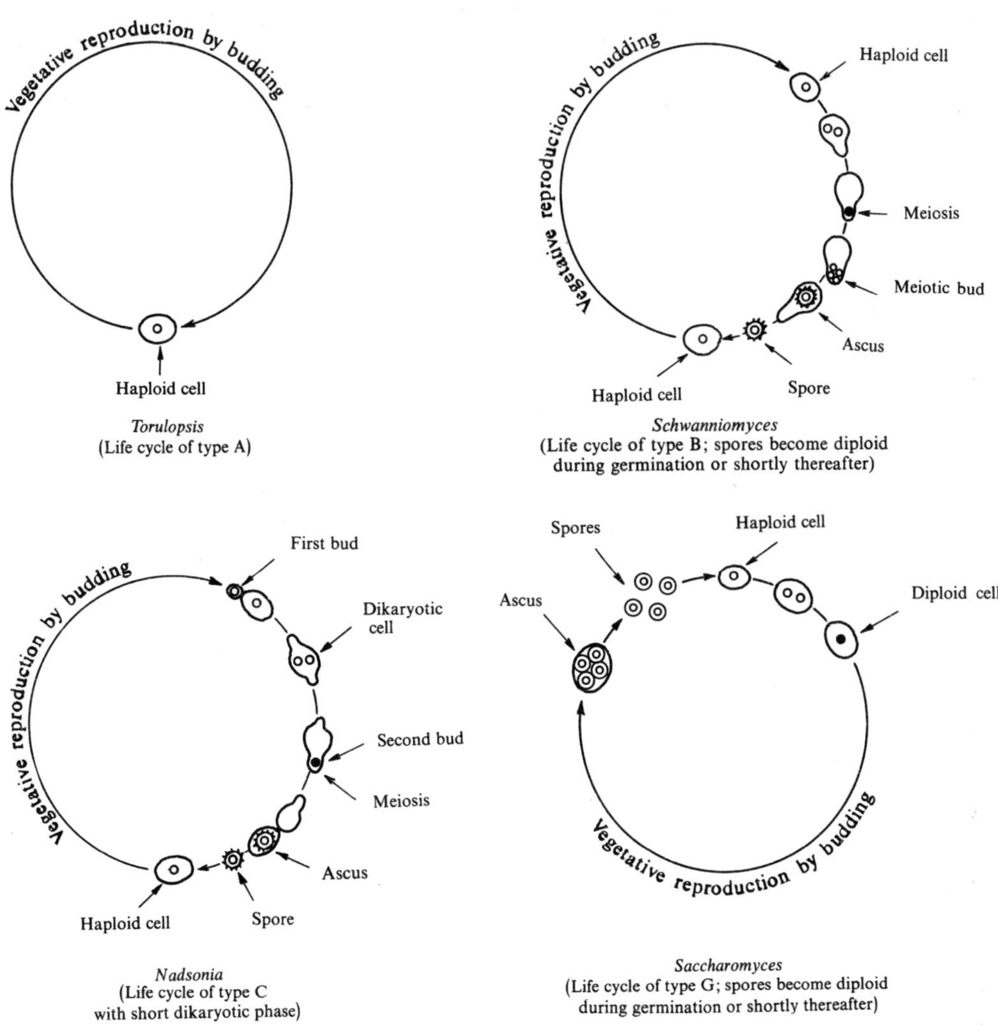

Fig. 74. The life cycle of some yeasts

led genetically by the dominant D gene. Diploidization takes place here immediately after spore formation and occasionally ascospores themselves may turn diploid.

Ascospores of *Saccharomyces ludwigii* turn diploid directly in the ascus while with other yeasts they become diploid only during germination. For instance, in *Saccharomyces cerevisiase* var. *ellipsoideus* the spore nuclei undergo mitosis and then the separated nuclei fuse during spore germination. It should be noted that owing to extended maintenance and transfers on stereotype media, some collection and brewery strains may lose their sporulation ability. The haploid phase is then completely missing.

A characteristic life cycle pattern of some yeasts is shown in Fig. 74.

Principal Features of Fungal Sexuality

Fungi usually reproduce sexually by fusion of two haploid nuclei, a process called karyogamy, which is sooner or later followed by meiosis. Karyogamy takes place either immediately after fusion of spores or hyphal fragments, or in specialized structures which are somewhat analogous to the female organs of higher organisms. The sexual mechanisms of fungi are distinguished not only by their morphology and cytology but also by their function.

According to their sexual behaviour, fungi are divided into homothallic and heterothallic types. Homothallic fungi are characterized by the fact that karyogamy takes place between nuclei which are sexually identical. In most heterothallic fungi fusion occurs between two sexually distinct nuclei. The differences between the nuclei are mostly those in the mating type. Heterothallism was discovered in 1904 by Blakeslee in the family *Mucoraceae*.

For the genetics of fungi the difference between homothallic and heterothallic fungi is of utmost importance. With homothallic fungi it is not possible to employ the technique of controlled sexual hybridization as the whole life cycle, including the diplophase, can be accomplished by starting from a single ascospore or conidium. All cell nuclei of homothallic fungi are compatible with respect to their mating type. Recombination can be achieved in some homothallic fungi solely by parasexual hybridization.

On the other hand, the full life cycle of heterothallic fungi can take place only if the fusing nuclei are sexually distinct. Heterothallic fungi thus contain two types of nuclei which differ in their mating type. Nuclei of the same mating type are mutually incompatible and cannot fuse. If a monosporial culture is started from a single conidium or ascospore, a large cell population will develop because of the large number of mitoses. All cells of the heterothallic fungus remain haploid throughout the life of the colony as the cells multiply only vegetatively and their nuclei have exactly identical genotypes. The mutual incompatibility of the mating types prevents the initiation of the sexual cycle in the particular clone.

The mating type is genetically fixed; with most heterothallic ascomycetes it is carried by a single locus and in most heterothallic basidiomycetes by two loci. The mating type is usually designated in bipolar species with (+) and (−), or **A** and **a**, or *a* and α. In such fungi a controlled sexual mating (crossing) can be accomplished.

Principles of Mendelian Heredity in Fungi

It should be known by now that all characters controlled by genes positioned in the chromosomes are subject to Mendelian inheritance. The expression of nuclear genes in the phenotype is thus closely connected with the behaviour and distribution of chromosomes during the cell cycle.

Of the various phases of fungal life cycles karyogamy and meiosis are of paramount importance for studying Mendelian inheritance. During karyogamy when two haplid nuclei fuse, the two sets of parent chromosomes are combined. During meiosis the individual genes are segregated and occasionally the integrated genes may be recombined.

Haploid organisms, especially unicellular and monokaryotic ones, are suitable for applications of mutagens and for isolation of mutants, as well as for obtaining a genotypically pure culture. The haploid character is present in most spores, such as conidia and ascospores. A suspension of such spores can be used for studying mutagenesis and for obtaining physiological (e.g., biochemical) mutants.

The course of nuclear division, i.e., mitosis and meiosis, cannot be observed under the microscope in most fungi. One of the exceptions to this rule is *Neurospora crassa*, where the position of the dividing nucleus can be followed optically in both stages of meiosis. This fungus was found to contain a haploid number of chromosomes ($n = 7$) in its cell nuclei while a double number of chromosomes was found during the diplophase or in a zygote ($2n = 14$). In all instances observed, the consequences of mitosis and meiosis in fungi resemble those in higher organisms.

The products of meiosis in fungi usually become nuclei of reproductive cells; in the *Ascomycetes* they appear directly as nuclei of the ascospores and in the *Basidiomycetes* they appear as nuclei of the basidiospores (or ballistospores in some yeasts) or of the sporangiospores in the case of *Mucorales*.

Tetrad Analysis

Whereas in most biological objects it is not technically possible to examine the four direct products of a given meiotic division, the possibility exists with some fungi. This is especially true for fungi of the class *Ascomycetes*, where the four principal products of a meiotic division remain together in a special cellular structure, the ascus. These four products of meiosis, designated a tetrad, are the basis of nuclei of the future ascospores. In most ascomycetes the tetrad is subsequently doubled or reduced. For example, the eight-spore ascus of such a fungus is the result of a mitotic division of the original four products of meiosis.

As mentioned before, ascospores are normally haploid. This fact makes it possible to follow the results of hybridization by direct genetic analysis of the individual tetrad members. This can be done optically directly, but biochemically only after multiplication of the individual isolated spores. In haploid ascospores the segregation of the individual genes is expressed irrespectively of their dominant or recessive nature, even in the case of intrachromosomal recombination of linked genes after crossing over. Tetrad analysis has the advantage that it is possible to determine whether any two genes are independently assorted or whether they are linked, without the necessity for further crossing.

The organization of the individual spores in an ascus is characteristic of the individual species of ascomycetes; basically, the tetrads can be ordered or unordered.

Ordered tetrads, which are particularly suitable for genetic studies, occur mainly in the genera *Neurospora, Sordaria* and *Podospora.* In these ascomycetes the definite sequence of all the four products of meiosis in determined by a fixed position of the division spindles in the two stages of meiotic division, both heterotypic and homoeotypic. Analysis of the ordered tetrads indicates whether the segregation of the genes (alleles) examined took place during the first (heterotypic) or second (homeotypic) meiotic division.

The situation is illustrated in Fig. 75, where the fate of a part of the nuclear material of a diploid ascogenic cell symbolized by one pair of chromosomes is shown. The chromosome pair may carry the information for ascospore colour: allele **A** codes for a black and allele **a** for white colour of spores. After meiotic division and subsequent mitosis the original diploid cell gives rise to an ascus with eight spores coloured white and black in a 1 : 1 ratio. The sequence of these spores is determined and corresponds to the course of the meiotic division. If no crossing-over took place during the meiotic division of the nucleus between the centromere and the pair of alleles investigated, allele **A** separated from allele **a** during the first meiotic division (upper part of Fig. 75 A). If, however, crossing-over did take place between the centromere and the pair of alleles during meiosis, allele **A** separated from allele **a** only during the second meotic division. This gave rise to a new arrangement of spores in the ascus (lower part of Fig. 75 B).

In other species of ascomycetes, such as many ascosporogenic yeasts or *Aspergillus nidulans,* unordered tetrads are generally found. This lack of order is again due to the course of meiotic division when the spindles are not regularly oriented in space. From the sequence of spores in the ascus the course of reductive division cannot be assessed as it cannot be decided whether the dominant and the recessive alleles of a given pair separated during the first or during the second meiotic division. The orientation of the two alleles with respect to the centromere can no longer be tracked; the two sister chromatids were kept together during the first meiotic division and could separate only during the second division.

In carrying out tetrad analysis a large number of asci must be treated and evaluated. Tetrad analysis makes it possible to construct genetic maps of chromosomes even with species where the chromosome structures cannot be observed under the microscope.

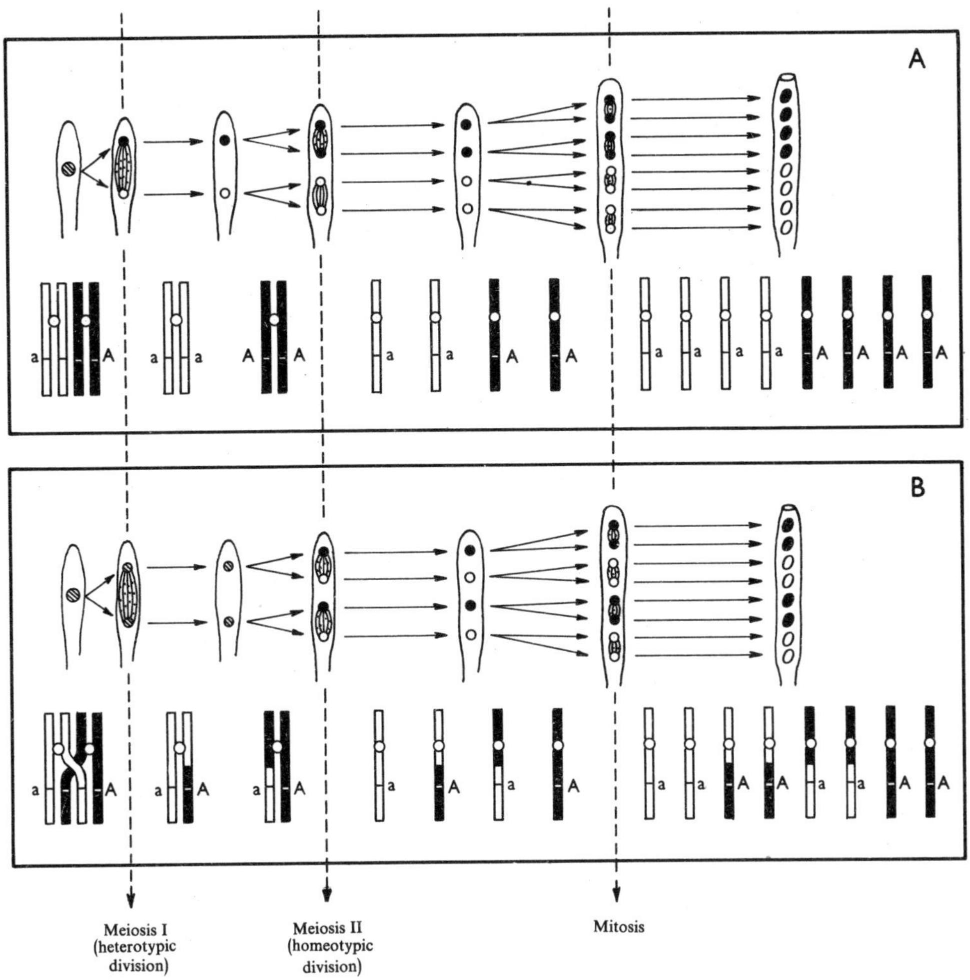

Fig. 75. An illustration of what happens to a pair of homologous chromosomes during the formation of an eight-spore ascus.
A, Without crossing-over; B, with crossing-over

Hybridization of Heterothallic Fungi

Neurospora crassa

The heterothallic ascomycete *Neurospora crassa* is probably the genetically best understood fungus. Because of its heterothallism it can be crossed by methods used with higher organisms. It is an ideal object for studying segregation and gene linkage, particularly with the ordered type of tetrads. It has often been used as a tool in biochemical genetics.

The vegetative phase of the life cycle of *Neurospora crassa* is haploid and appears as a freely branching multicellular mycelium. The cells of this mycelium are polykaryotic. The septa between these cells are perforated and allow for the migration of nuclei through the mycelium. This haploid mycelium is able to form two types of sexual spores – monokaryotic microconidia 1–4 μm in diameter, and a small number of polykaryotic macroconidia 6–12 μm in diameter. However, some strains are known to produce only microconidia. The mycelium of this fungus may be propagated using hyphal fragments. The mycelium and all the spores derived therefrom occur in two mutually compatible mating types, which are designated **A** or **a**. This genetically fixed existence of two mating types is at the basis of heterothallism. In haploid cells heterothallism is controlled by a single allele. Alleles determining a given mating type are stable. Mutations that would result in a loss or an acquisition of the mating type have never been observed. The vegetative stages of the two mating types are hermaphroditic, as every type can act during crossing as a male or as a female germ cell.

Older mycelia of both mating types are made of twisted hyphas to form nidiform organs called protoperithecia, which are the true female sexual organs. Special hyphae, called trichogynes, grow out of the protoperithecium and trap conidia of the opposite mating type. No special male organs or male gametes are formed here. Male nuclei can be donated by conidia or by vegetative hyphae of the opposite mating type. The male cell nucleus of the opposite mating type then migrates through the trichogyne and initiates the dikaryotic phase in the ascogenic cell of the protoperithecium. In the ascogenic cell the male and female nuclei undergo mitosis (conjugative division) which gives rise to a large number of dikaryotic cells. Terminal dikaryotic cells of ascogenic hyphae then form the basis for the ascus. Nuclear fusion in these terminal dikaryotic cells produces zygotes. The diploid nuclei of the zygotes undergo meiosis, which generates four haploid nuclei. Subsequent division of these four nuclei then gives rise to eight haploid nuclei of the future ascospores, every two of the nuclei being identical (details of ascus formation are shown in Figs. 71 and 72). In the course of development the cavity of the protoperithecium increases, acquires its typical shape during ascus maturation and changes into the perithecium.

Every perithecium can contain a greater number of asci (2 to 300), each of which is derived from the pair of nuclei which fused during karyogamy. In *Neurospora crassa* the ascospores in every ascus are ordered according to the sequence of nuclear division during meiosis.

Under optimal cultivation conditions the entire sexual cycle of *Neurospora crassa* takes 2 weeks. Then the ascospores require a further 2 weeks of rest before they can germinate quantitatively (usually after exposing them to a temperature shock of 30–45 min at 60 °C).

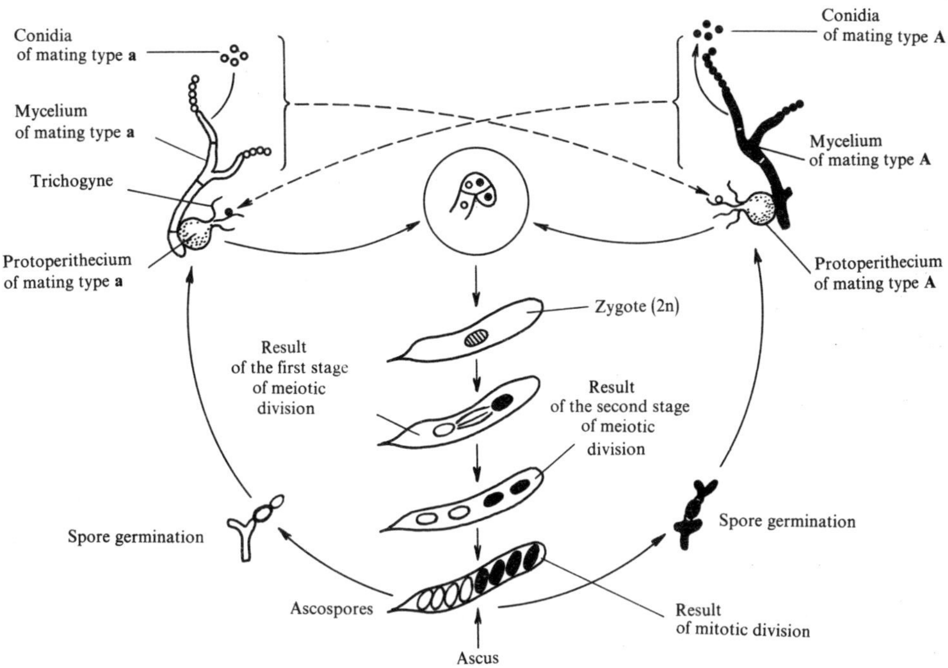

Fig. 76. Life cycle of the heterothallic fungus *Neurospora crassa*

The course of the life cycle of *Neurospora crassa* is shown schematically in Fig. 76, where the pairing types are distinguished as black and white.

Let us now undertake a model hybridization experiment in which tetrad analysis will be used to follow the gene responsible for the colour of ascospores. Let us assume that the allele for this property contained in a chromosome of one mating type of *Neurospora crassa* (mating type **A**) is the one coding for black colour, the allele in an analogous chromosome of cells of the opposite mating type (**a**) then codes for white colour. Monosporial cultures are first prepared from each of the parent strains differing in both the mating type and the gene for ascospore colour. The sexual cycle starts and the asci in the perithecium begin to develop only on mixing of these two cultures. As the tetrad is ordered in this fungus the asci formed will be of six types differing in the sequence of black and white ascospores (Fig. 77).

If the allele for ascospore colour is placed in the chromosome near the centromere or if the crossing-over process is prevented in some other way asci of types I and II will be formed. If a crossing-over event takes place during the meiotic division allele **A** is separated from allele **a** only during the second stage of meiotic division and the asci are then of types III to VI (cf., Fig. 75).

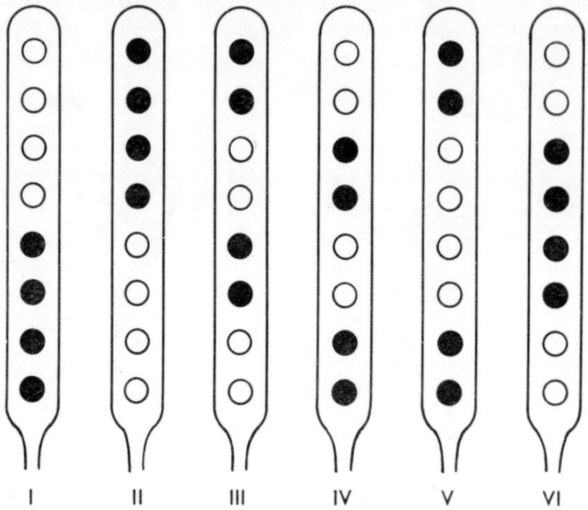

Fig. 77. Types of asci with ordered tetrads when following two alleles

The probability that a pair of homologous chromosomes will form a chiasma between the centromere and a given gene pair (and hence a crossing-over will be possible) increases with the distance of the locus examined from the centromere. The fact that the percentage of occurrence of asci with a "recombined" sequence of ascospores reflects quantitatively the distance of the given gene from the centromere offers possibilities for constructing a chromosome map.

If two or more pairs of alleles are examined, the genetic analysis of ordered tetrads makes it possible to determine (a) whether the allele pairs examined are independently assorted or linked and (b) the relative distance between individual linked genes and the centromere.

If the crossed partners differ in a greater number of characters the number of theoretically possible tetrad types increases exponentially. While six possible ascus types existed after crossing individuals differing in a single character, 36 possible types were encountered when examining dihybrid crosses.

Saccharomyces cerevisiae

The greatest effort of yeast geneticists has been expended on the species of baker's yeast *Saccharomyces cerevisiae*. This is a heterothallic yeast, its mating type being determined by a single pair of alleles. Heterothallism of yeasts was discovered in 1918 by Kruis and Šatava.

Saccharomyces cerevisiae can reproduce asexually by budding, which follows a mitotic division of the nucleus. During budding the genetic material of the chromosomes is evenly distributed between the mother cell and the bud but the cytoplasmic material is not distributed indentically. This gives rise in the yeast to a vegetative phase of the life cycle which under certain conditions can last indefinitely. Under normal conditions of vegetative growth and reproduction the cells of *Saccharomyces cerevisiae* are diploid*, but if an ascosporogenic yeast is cultivated on a suitable medium (such as acetate agar) sporulation sets in after a meiotic division of the diploid nucleus. The four haploid products of this meiotic division represent the nuclei of the future ascospores, which then fuse to a diploid zygote. The complete life cycle of *Saccharomyces cerevisiae* then consists in an alternation between a haploid and a diploid nuclear phase, the haplophase being limited to the immediate products of meiosis, i.e., ascospores.

In the ascus the ascospores of *Saccharomyces cerevisiae* form unordered tetrads. The position of the spindles during the individual phases of meiotic division is irregular and hence the arrangement of the spores in the ascus is also random.

In genetic studies of a heterothallic yeast one proceeds from a hybridization experiment carried out with two haploid forms of yeast, i.e., ascospores or haploid vegetative cultures of opposite mating types and with different genotypes. A statistically significant number of asci are then subjected to a tetrad analysis. The technique in the yeast requires first an enzymic rupture of the ascus wall and then release of the spores. The spores are taken out using a micromanipulator. The individual spores of an ascus are transferred to separate droplets of a suitable medium and, after growth of the haploid culture from the isolated ascospores, their genotype is determined.

Genetic analysis of unordered tetrads does not provide such unequivocal information on gene positions as genetic analysis of ordered tetrads would (such as in *Neurospora crassa*). The random organization of spores in the ascus of the yeast *Saccharomyces cerevisiae* does not make it possible to determine whether the alleles examined were segregated during the first phase of meiotic division (heterotypic division) or in the second phase (homoeotypic division). The unordered nature of tetrads then prevents the use of the centromere as an auxiliary marker so that the sequence of spores in an unordered ascus does not indicate whether crossing-over has taken place between the centromere and the given gene.

If a single pair of alleles (**Aa**) is examined in the yeasts *Saccharomyces cerevisiae* using tetrad analysis one can only determine whether the segregation of these alleles is reciprocal (2**A** : 2**a**) or nonreciprocal (e.g., 3**A** : **a**). Ascus types with nonreciprocal segregation are rare and are formed as a consequence of certain abnormalities.

---

* In many yeast species polyploid forms are known, although their existence was established only genetically and not cytologically. Triploid yeasts result from a fusion of a haploid and a diploid cell, tetraploid yeasts arise from two diploid cells, etc.

When following segregation of two pairs of alleles **(AaBb)** three types of asci are formed from unordered tetrads. Two are bivalent (ditypes), which means that they have the same number of ascospores of two genotypes. One of these ditypes contains the parent genotypes (2**AB** : 2**ab**) and the other contains recombinants (2**Ab** : 2**aB**). The third type of ascus is tetravalent (tetratype) as each of its four ascospores has a different genotype **(AB** : **Ab** : **aB** : **ab)**. If three and more allele pairs are examined the number of distinct tetrad types increases exponentially.

The representation of the individual ascus types in a genetic analysis of unordered tetrads is determined (a) by the position of the given loci (in homologous or nonhomologous chromosomes), (b) by the distance between the given loci and the corresponding centromere, (c) if the loci are linked, by their mutual distance as well as by the distance from their common centromere (greater distances mutually as well as from the centromere result in higher frequencies of simple and multiple crossing-over events) and (d) if the loci are linked, by the type of linkage phase (coupling **AB/ab**, or repulsion **Ab/aB**).

Genetic analysis of unordered tetrads consists in the determination and evaluation of statistically significant frequencies of the individual tetrad types.

Tetrad analysis of unordered asci can provide certain information on the localization of genes even as two allele pairs are examined. The decisive factor here is the ratio of frequencies of the parent ditype (2**AB** : 2**ab**), of the recombined ditype (2**Ab** : 2**aB**) and of the tetratype **(AB** : **Ab** : **aB** : **ab).** If it is found that these types of asci are present in a ratio of 1 : 1 : 4, it indicates that the examined genes are independently assorted. However, it is not possible to decide whether the genes are localized on two different chromosomes. If, however, a greater frequency of parent than recombinant ditypes is found and if the frequency of the tetratypes is less than 2/3, the examined genes may be assumed to be linked. However, here again it is not possible to decide whether both pairs of alleles lie on the same chromosome or on different chromosomes close to their centromeres. The examples indicate that examination of two allele pairs by a genetic analysis of unordered tetrads does not provide unequivocal conclusions. Therefore, at least three markers are used in these analyses. One of the techniques uses the so-called centromere labelling, where a gene lying immediately at the centromere is used. The probability of crossing over between this gene and the centromere is practically zero so that the marker can substitute for the centromere in our considerations.

Hybridization of Homothallic Fungi

The most frequently used genetic model of homothallic fungi is *Aspergillus nidulans*. This mould belongs to the homothallic ascomycetes and its nutrition in the wild type is prototrophic.

Colonies of this organism are formed by a mesh of septated hyphae which gives rise to the mycelium. The hyphal cells are polykaryotic and the individual nuclei are

The nuclei of the dikaryotic cell of an ascogenic hypha divide mitotically and the tetrakaryotic cell formed bends in a beak-like fashion.

The terminal dikaryotic cell separates to form the future ascus.

Both vicinal monokaryotic cells fuse into a single dikaryotic cell.

Fusion of the nuclei in the tip cell
gives rise to a diploid zygote which forms the basis of the future ascus. At the same time, the process repeats itself as described in the first drawing of this schematic illustration. This then leads to the formation of another zygote and to the development of another ascus.

We shall now follow the fate of the first diploid zygote. It undergoes first a heterotypic meiotic division which is followed by a homeotypic one.

Subsequent mitotic division then gives rise to eight haploid nuclei

which are enveloped into membranous structures and give rise to ascospores.

Before maturation of the ascospores the nucleus undergoes another mitotic division so that every ascospore capable of germination contains two haploid nuclei.

Fig. 78. Development of an ascus in *Aspergillus nidulans*

haploid under normal conditions – a vegetative haploid state. In a certain developmental phase the surface of the colony is covered by aerial hyphae, conidiophores, which carry asexual spores, the so-called conidia. The conidia are monokaryotic and a row of conidia in the conidiophore is formed by a mitotic division of the original nucleus of a monokaryotic secondary sterigma. The sexual part of the life cycle of this homothallic fungus begins with the formation of protoperithecium from the hyphal agglomerate. The ascogenic layer of the protoperithecium is formed by dikaryotic cells with haploid nuclei. Fusion of these nuclei gives rise to the diploid nucleus of the zygote, which divides immediately through mitosis to four haploid nuclei. These nuclei then undergo mitosis and the eight nuclear products are at the basis of eight future ascospores (Fig. 78).

Asci of *Aspergillus nidulans* are unordered and contain eight spores each. Mature perithecium is a thick-walled spherical structure with the diameter of 200 – 400 μm, containing various quantities of asci with the diameter of 6.8 – 8.5 μm.

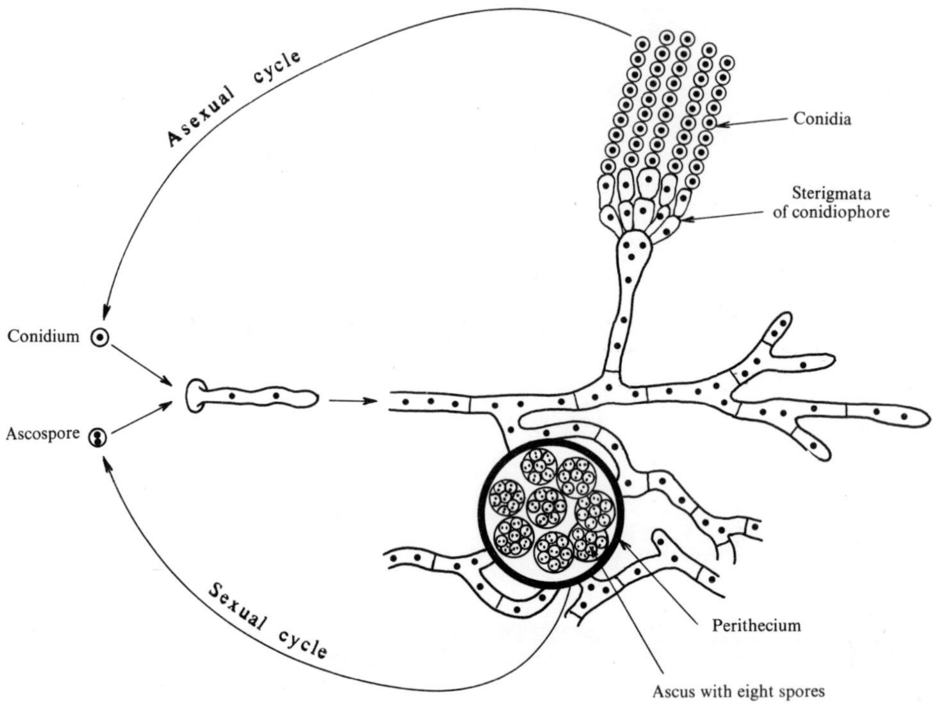

Fig. 79. Life cycle of the homothallic fungus *Aspergillus nidulans*

The life cycle of *Aspergillus nidulans* is shown in Fig. 79. Both Figs. 78 and 79 show that, as with other eukaryotic organisms, there is an alternation between the haploid and the diploid nuclear phases, the diploid phase being limited to the zygote under normal life conditions. Meiotic division of the zygote follows immediately after its formation and the ascospores thus formed are again haploid. *Aspergillus* is thus a fungus with a haploid vegetative stage. This species distinguishes itself by another rare feature: its haploid nuclei in the hyphae may occasionally fuse, the process occurring spontaneously. The diploid nuclei then undergo mitosis; if conidia are formed they are monokaryotic but diploid. They are larger than normal (haploid) conidia. Perithecia formed from the mycelium of these diploid strains contain 16 spores per ascus. Fusion of nuclei in hyphal cells offers another exceptional possibility of recombination, through parasexual type hybridization.

The parasexual cycle is based on several phenomena that will be discussed now.

Heterokaryosis

If a cell contains more than one nucleus, the nuclei differing in their genotype, the cell is heterokaryotic and the situation is heterokaryosis. The phenotypic expression of a heterokaryotic cell is determined by an integration of all the genotypes of the haploid nuclei present.

As mentioned before, the cells of *Aspergillus nidulans* are polykaryotic. The hyphae which are coiled and twisted in the mycelium form occasional anastomoses so that cells belonging to different hyphae fuse with each other. If conidia of two genetically

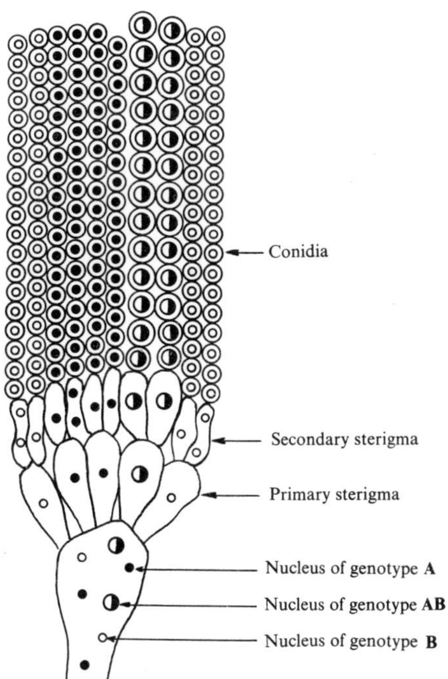

Fig. 80. Origin of conidia with different karyotypes in *Aspergillus nidulans*

distinct strains are cultivated together, the resulting culture will contain hyphal cells of both genotypes. This again is a heterokaryotic mycelium. In view of the fact that nuclei of heterokaryotic cells retain their individuality, the heterokaryon dissociates again during formation of conidia; each chain of conidia then contains the genotype which corresponds to the nucleus contained in the secondary sterigma (Fig. 80).

In a perithecium which is formed on a heterokaryotic mycelium any two nuclei may fuse as follows from the homothallic character of this species. If the genotypes of the two parental types are designated **A** and **B**, three types of zygotes, **AA**, **BB** and **AB**, may arise. Asci with genotype **AA** or **BB** will yield genetically uniform ascospores in the entire parathecium. A possibility of genetic recombination exists only with zygotes of type **AB**, where meiotic crossing-over can take place during ascospore development, with the genetic consequences already described above. This is an example of true hybridization taking place during the sexual process.

Parasexual Cycle

The basic prerequisite of a parasexual cycle is the preparation of the so-called balanced heterokaryon, which is characterized by the presence in it of two genotypically distinct nuclei which provide for mutual nutritional complementation. A balanced heterokaryon can exist indefinitely and possesses a selection advantage over any of the parent strains.

Irradiation of a suspension of conidia of a prototrophic strain of *Aspergillus nidulans* yields two different auxotrophic mutants; one of them requires riboflavin (B_2^-) and the other requires lysine (Lys^-) for growth. The first genotype is then $B_2^-Lys^+$ and the second is $B_2^+Lys^-$. Stable heterokaryons with a high content of heterokaryotic hyphae are obtained from a thick inoculation of a mixture of spores of the two strains on a nutrient medium on which none of the strains will grow but on which it will germinate. As soon as the hyphae of germinating spores fuse, a heterokaryon of the genotype $B_2^-Lys^+ + B_2^+Lys^-$ is formed and this can thrive on the nutrient medium used. Conidia will be formed in the heterokaryotic mycelium and will develop into the original auxotrophic strains – the heterokaryon dissociates into its original components during conidiation.

As mentioned before, *Aspergillus nidulans* can undergo diploidization of mycelial cells*. This can even occur in the heterokaryotic cells of the mycelium; if two haploid nuclei of distinct genotypes fuse, a heterozygous diploid nucleus is formed and this can be the basis of a diploid culture with a $B_2^-B_2^+Lys^-Lys^+$ genotype. The conidia of this strain will also be diploid and possess the same genotype.

Formation of a heterozygous diploid represents another possibility of achieving gene recombination in homothallic fungi outside the sexual mode. The nuclei of these

---

* An increase in diploidization frequency can be achieved in fungi by exposure of cells to vapours of camphor.

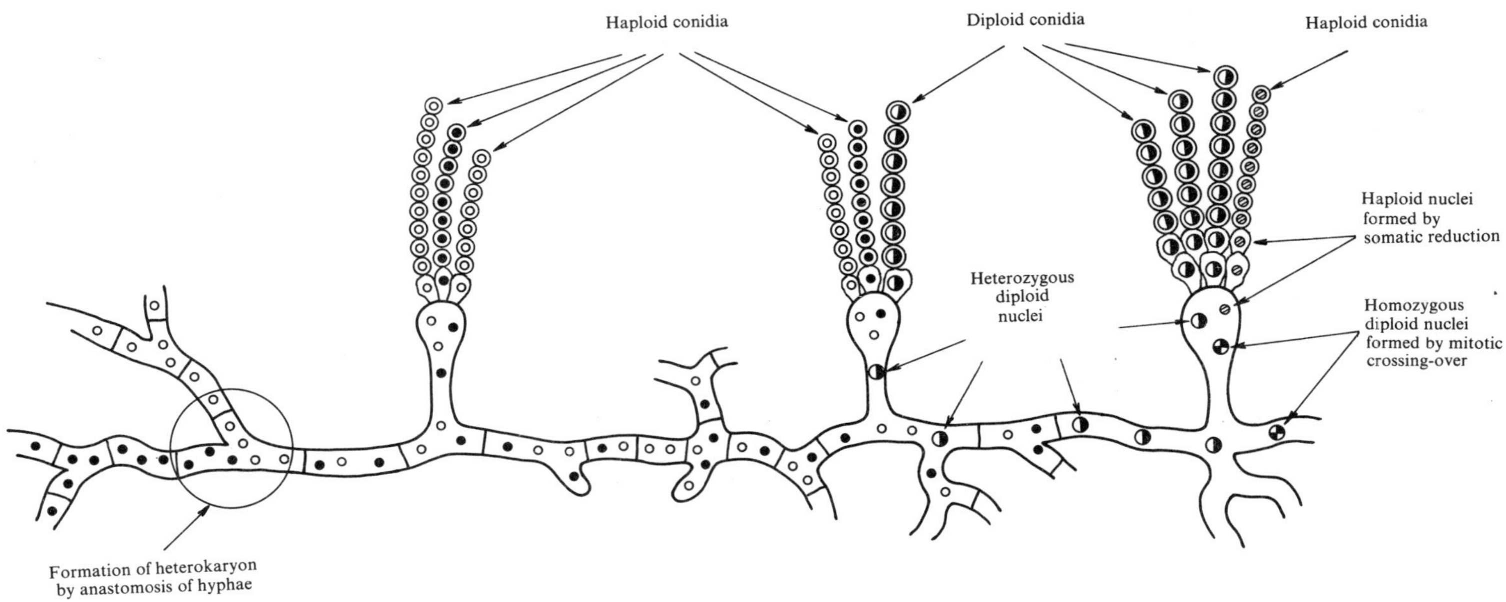

Fig. 81. Karyological changes during the parasexual cycle of *Aspergillus nidulans*

diploid cultures can undergo two other genetically important changes at a certain low frequency, viz., (1) mitotic crossing-over and (2) somatic reduction (haploidization).

The mitotic crossing-over is equivalent to a recombination of the genetic order during mitosis rather than during meiosis as is usual. Homozygous diploid nuclei can then be formed. The frequency of genotypes in diploid organisms can serve for mapping the position of linked loci with respect to the centromere of the chromosome on which they are located.

Somatic reduction decreases the diploid number of chromosomes to the haploid one, not through meiosis but through what is called nondisjunction (i.e., irregular distribution of chromatids). Recombination of linked genes does not take place as somatic reduction affects entire chromosomes. The frequency of various genotypes in the haploids formed can be used for determining the localization of the examined genes either in homologous or in heterologous chromosomes.

Fig. 81 illustrates the karyological changes occurring during the parasexual cycle of *Aspergillus nidulans*.

The parasexual cycle was also observed in *Aspergillus niger, A. sojae, A. oryzae, Penicillium chrysogenum* and *Fusarium oxysporum*. It should be noted that *Aspergillus niger* and *Penicillium chrysogenum* exhibit no sexual cycle under laboratory conditions. The formation of recombinants in these fungi is apparently solely dependent on the parasexual cycle.

Parasexual cycles were detected even in a number of streptomycetes, even if with certain differences.

Mitotic crossing-over is apparently not a speciality of fungi – it has been found even in *Drosophila*. Mitotic recombination is an important process, not only from a theoretical but also from a practical point of view; it makes it possible to carry out genetic analysis and controlled breeding of organisms without a sexual cycle. Mitotic crossing-over can occur even in tissue cultures of higher diploid organisms, such as of human cells.

Appendix

Molecular Evolution of Living Matter

The universe is thought by many to have been created by a "Big Bang" about 18 billion years ago. About 4.5 billion years ago the Earth appeared on the scene, the event having been apparently preceded by a stage of smaller bodies, planetesimals which had formed by condensation of interstellar dust and gas. The planetesimals "fused" within a cosmically very brief period of about 1 000 years into the protoplanets, evolutionary stages of all planets of the solar system. The formation of the Earth's crust can be timed with relative accuracy to 3.8 billion years ago. The primordial atmosphere on the Earth had a reducing character, containing nitrogen, hydrogen, methane, ammonia and water vapour. The hydrogen soon disappeared from the atmosphere and was replaced by oxygen-containing components, especially carbon monoxide. It is generally assumed that during the origin of life the atmosphere did not contain oxygen; oxygen appeared later as a product of photosynthesis.

Formation of Organic Compounds

There is enough evidence now for the view that the so-called chemical evolution could have started by reactions of the simple components of the primeval atmosphere and of inorganic components of the geosphere. The basic building block for the construction of organic compounds under abiotic conditions is believed to be hydrogen cyanide, which is formed by the following two reactions:

$$2\,CH_4 + N_2 \rightarrow 2\,HCN + H_2$$

$$CO + NH_3 \rightarrow HCN + H_2O$$

These reactions can be initiated by heat (of volcanic origin), by UV radiation from the Sun or by electrical discharges (lightning of atmospheric origin).

Hydrogen cyanide, in its turn, can give rise under the primitive prebiotic conditions to a number of reactive compounds which, through further reactions, can yield the building material for living matter:

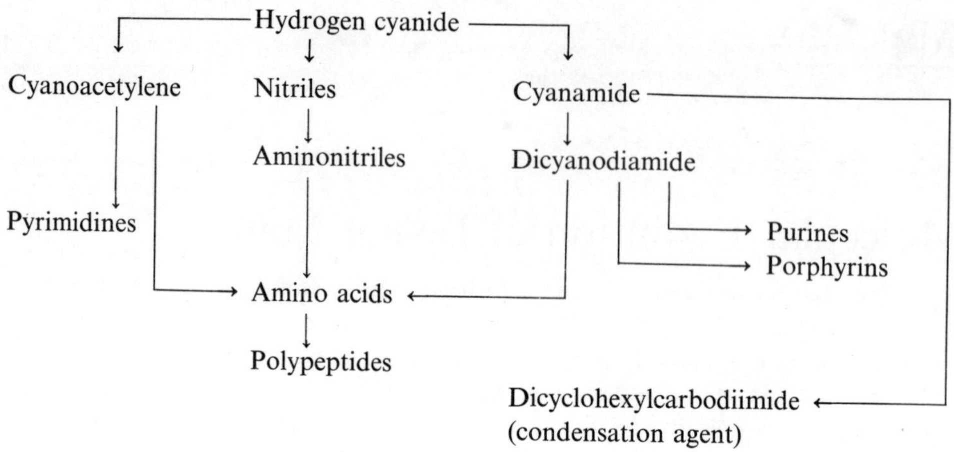

Amino acids could only be formed in a reducing atmosphere and, therefore, they cannot arise by the above process under present-day conditions on Earth.

Under the conditions of the primitive Earth there was a good chance of formaldehyde being formed, which, in the presence of limestone, can produce various sugar derivatives.

The Earth was a hostile place during chemical evolution. Pouring rain accompanied by thunderstorms provided the milieu which, aided by UV light arriving from the Sun, activated the reactions of simple molecules. The organic substances formed could accumulate in the primeval sea as there were no microorganisms or enzymes that would destroy them. This period may have lasted for about half a billion years. During this time all the amino acids, pyrimidines, purines, sugars, fatty acids and other principal building blocks of living matter were synthesized. The hot ocean containing these compounds is often called the primeval soup, as the concentration of dissolved organic compounds may have locally reached as much as 10%.

This view is not completely speculative, as experiments can be carried out to show that exposure of a gaseous mixture of simple compounds probably occurring in the primitive atmosphere to UV and visible light, X-rays, γ-radiation, silent or spark discharges, ultrasound, pressure waves or streams of α and β particles can give rise to several hundred organic compounds, among which all the important groups of living matter constitutents are present.

Organic compounds were apparently formed at other places in the universe, as documented by their presence in extraterrestrial bodies and their fragments that have reached the Earth. Some meteorites were found to contain amino acids just like those in present-day proteins, as well as some others not known on Earth. However, these amino acids are optically inactive D,L-forms. Even in the more distant universe, in interstellar clouds, a number of organic compounds have been detected, such as formaldehyde, carbon monoxide, hydrogen cyanide, methanol, formic acid, acetaldehyde and a number of others. Even the dust covering the lunar surface was found to contain traces of amino acid precursors.

Formation of Polymeric Compounds

It was seen that energy was required for the formation of the fundamental building blocks of living matter. The next step in the evolution of organic compounds was the formation of covalent bonds between the simple blocks which gave rise to polymers. The formation of polynucleotides and polypeptides required either an elevated temperature (130 – 180 °C) or the presence of some condensation agents. These agents, such as dicyclohexylcarbodiimide, could be formed from cyanamide. Polymers could also be generated on the surface of some minerals, such as montmorillonite. Long polypeptides are formed "artificially" in this way in high yields. However, the most likely route to the formation of polymers was that catalysed by polyphosphates and their esters, which even now are among the most potent biological condensation agents. Moreover, the formation of polyphosphates under prebiotic conditions may have proceeded with ease. Polyphosphates react sluggishly with water and are thus kinetically fairly stable. At the same time, they are unstable thermodynamically, their hydrolysis being characterized by a large negative change of the Gibbs free energy, $\Delta G^{\circ\prime}$. This combination of properties makes ATP a suitable condensation agent. The adenosine moiety of ATP represents the link for attaching ATP to the enzymes.

Elevated temperature or the presence of condensation agents thus provided the conditions necessary for the formation of polypeptides of high-molecular weight which possessed some properties of proteins. These proteinoids even exhibit certain catalytic properties resembling primitive enzymes: they increase the rates of ester hydrolysis, decarboxylation of α-oxo acids and of deaminations. However, further development of proteinoids towards present-day enzymes required the presence of a template which would ensure the correct sequence of amino acids.

By analogy, polynucleotides could be formed by condensation of nucleotides. Here, however, the polynucleotides themselves could act as templates as they are able to pair with other polynucleotides through hydrogen bonds. It is known, for instance, that AMP polymerizes much more rapidly in the presence of poly(U) than without it.

With a few exceptions, biological systems contain only L-isomers of amino acids. According to modern views, the L-amino acids were selected as building blocks rather accidentally (D-amino acids would have served equally well). Once they had been selected, the selection of L-forms had to be maintained as a stable α-helix of a protein chain cannot contain both L- and D-forms of amino acids. Likewise, during nonenzymic condensations of polynucleotides the stereospecificity of the template determines the steric form of the incorporated monomers.

To summarize, all the principal components of living matter could have been formed by a strictly organic chemical evolution with an input of energy. The earliest organic material was found in Precambrian rocks about 3.1 billion years old. These strata contained isoprenoids, porphyrins, purines and pyrimidines, as well as sterol-like hopanoids. This would indicate that life may have been around for nearly 3.1 billion years. Interestingly, many of the rocks examined from that period were found to contain "microfossils" resembling present-day bacteria and even yeast cells, with diameters of about 1 μm and over 10 μm, respectively.

Origin of Complex Systems

In the course of chemical evolution the primeval sea contained certain types of organic molecules, particularly the stable ones. These molecules were likely to form complexes which were thermodynamically more stable than their components. Every step of chemical evolution was apparently facilitated by the fact that the entropy of the system and of its aqueous environment increased. It may even be assumed that several complexes associated and acquired a certain functionality, such as catalytic activity or an ability to counteract changes in the surroundings. It cannot be excluded that an important role was played here by the ability to form primitive templates required for replication.

It is extremely probable that such organized matter formed at many places on the Earth and over millions of years over and over again, in different modifications and with different degrees of complexity. However, a single "line" overcame the others and formed the foundation for the future evolution of living forms. More than 99% of organisms that have existed on Earth have become extinct, leaving no progeny. By analogy, it may be conjectured that prebiotic evolution proceeded with strong selectivity which removed all the forms of incipient life that did not possess the required survival potency.

The first structures with the attribute of life obviously were extremely primitive by present-day standards and their only function was to perperuate their own existence. These structures could certainly exist even without any genetic material.

The discussion of whether proteins or nucleic acids were here first has been going on for decades. Three answers are theoretically possible. According to the first, the primeval "cell" could survive with just one type of primitive catalyst, a polypeptide, the genetic material having arrived subsequently. The second theory assumes the primary role of nucleic acids with proteins arriving later, according to the information stored in the nucleic acids. According to the third view, cells could only be formed from complexes of nucleic acids with proteins.

In this context, Muller's hypothesis should be mentioned, according to which the principal and decisive prerequisite of life is the synthesis of a gene under prebiotic conditions; this prerequisite is met by nucleic acids as they can replicate themselves, they can code for proteins and can undergo mutations. This theory is supported by the great variety of enzymes that contain a nucleotide as a coenzyme. It is likely that the first informational macromolecule was RNA rather than DNA. What is still unexplained is how the genetic code and the complex protein-synthesizing machinery evolved.

Muller's hypothesis is supported by the existence and type of replication of some viruses and phages. Thus, tobacco mosaic virus can be transmitted simply by RNA isolated from it. The question of the extent to which phages and viruses took part in the development of living systems is still unanswered. It cannot be excluded that at a certain stage of evolution phage and viral particles represented important horizontally acting mediators between vertically evolving lines.

Formation of the Cell

The first consistent theory of the origin of the cell is that of Oparin. According to this theory, the first to be formed were droplets called coacervates, the generation of which can be easily simulated in the laboratory. In aqueous solutions of highly hydrated polymers part of the aqueous solution is enclosed into spherical membranes in a process called coacervation. As coacervates were formed, some catalysts could be enclosed in the droplets and these could mediate the translocation of substances from the surrounding medium, catalyse their "metabolism", cause volume increases of the droplets (i.e., growth) and even make the droplets divide (i.e., reproduce). Several types of interesting coacervates were made in Oparin's laboratory, some of them exhibiting a primitive metabolism. For these primitive cell precursors Oparin coined the term protobionts.

Fox and co-workers prepared interesting structures which they called proteinoid microspheres. They were formed under definite conditions from hot, thermally prepared, solutions of proteinoids. The microspheres possess a number of properties common with a real cell: they can bud in analogy to yeast cells; the buds separate from the mother microsphere and, together with a daughter population, they give rise to a new generation of microspheres which can go on budding.

On using basic proteinoids the microspheres will accumulate poly(C) and poly(U) but not poly(A) or poly(G). If these proteinoid-polynucleotide complexes are incubated with aminoacyladenylates, polypeptides are formed; these microspheres thus exhibit a catalytic function analogous to that of the ribosomes.

Development of Aerobic Living Systems

It is now generally believed that the first cells were anaerobic heterotrophs making use of organic substances in the primitive sea to construct their polymers as well as to "burn" them to obtain some energy from the process. These organic substances were gradually depleted from the environment and only those cells could survive which were able to utilize simple carbon compounds, such as carbon monoxide, as material and solar radiation as energy. This may have given rise to the first photosynthetic organisms, some 3 billion years ago. It is probable that these cells were not yet able to release molecular oxygen as they used hydrogen sulphide as a source of hydrogen, in common with the present-day sulphur bacteria. It was only much later that oxygenic photosynthesis appeared on Earth, probably effected by some forerunners of present-day cyanobacteria.

These photosynthetic bacteria kept producing oxygen and its concentration in the atmosphere gradually rose, very slowly at first. After the first billion years of photosynthesis the oxygen content of the atmosphere was about 1% and after the next 400 million years it was 10%. Higher plants and aerobic animals could evolve only about 2 billion years after the first oxygen-producing photosynthetic bacteria had begun to function.

The first aerobic cells utilizing oxygen had the enormous advantage over the anaerobes that they could obtain much more energy from a molecule of exogenous substrate (such as glucose). The formation of respiratory systems and oxidative phosphorylation brought the evolution of energy-converting metabolic processes to the present-day high level.

Comparative studies carried out with 16S ribosomal RNA brought about a kind of revolution in bacterial taxonomy. It was found that all the existing bacterial species must have had a common ancestor more than 3 billion years ago. The oldest genotypes are those of anaerobic bacteria, followed closely by photosynthetic bacteria.

An important milestone in the evolution of life was the formation of a eukaryotic cell. It is commonly believed that eukaryotic cells evolved as a consequence of metabolic endosymbiosis. The modern paragons of such an endosymbiosis are mitochondria and chloroplasts. Eukaryotic cells are larger and better equipped for contact with the environment than prokaryotic cells are. They could also differentiate more productively owing to their much larger genome. During biological evolution nonlethal mutants could be formed and selection then favoured those which were better equipped for the existing living conditions.

The origin of man is now placed in a period between 1 and 2 million years ago. If the history of our planet is likened to a 24-h day, then on that time scale man appears on Earth about 30 seconds before midnight.

Evolution of Nucleic Acids and Proteins

The character of self-replicating polymers was acquired by polynucleotides primarily owing to the unique and specific pairing ability of nucleic acid bases. Replication of a given polynucleotide based on the principle of base pairing necessarily requires the presence of a complementary chain. It is only the two polynucleotide chains or strands that can form the double helix of complementary segments. This fundamentally new quality made it possible for replicating molecules to effect evolutionary changes not only in the linear sense but also in their two- and three-dimensional structure.

Segments of nucleic acids displaying pair symmetry as are often encountered in nucleic acids were probably formed in such a way that the nucleotide copied itself going back from the end, this giving rise to the well known hairpins of different length.

Double-helical DNA known today may have appeared only later as a consequence of a selection process of the environment exerted not on the polynucleotide itself but rather on the product of its genetic expression.

Gene duplication may be counted among the more important evolutionary principles such as accompanied the development of haemoglobin in animals and of ferredoxin in bacteria. Reduplication or multiple amplification of a gene is extremely important from the point of view of evolution. While a biological function is guaranteed after amplification by a single gene, its "sister" genes are free for any type of innovation. Gene amplification opened the way for the evolutionary mechanism called exon shifting. Furthermore, "amputations" of certain parts of the central DNA regions may be observed as they took place in the process of evolution. Another important mechanism of molecular evolution undoubtedly was mutations causing substitutions of amino acids in proteins. Some such substitutions may have been advantageous whereas others were harmful for the host. A case in point is the consequence of a single point mutation in the gene for haemoglobin; the mutation resulted in the replacement of glutamic acid with valine in the β-chain of haemoglobin. The new type of haemoglobin, called haemoglobin S, is the cause underlying the appearance of a "molecular" disease, the sickle-cell anaemia which is heritable according to Mendel's laws. Heterozygous individuals who carry only one faulty allele for haemoglobin carry on a completely normal life while homozygotes with both faulty alleles only rarely reach the age of 30. However, individuals with sickle-cell anaemia receive a bonus with the genetic defect – they are resistant to the malaria plasmodium.

Some groups of proteins show such similarities that a common ancestor for all is evident. Thus, myoglobin and haemoglobin are probably derived from a common polypeptide, and so are trypsin and chymotrypsin. A striking similarity is found even between some proteins fulfilling very different biological functions, such as the milk protein α-lactalbumin and lysozyme isolated from egg white. Another example is the sequence homology of insulin with the neurotrophic hormone.

Molecular Clock

Comparisons of protein and DNA structures of present-day species make it possible not only to determine their evolutionary paths but also the period when their development diverged. Chromosomes thus carry a kind of code of the temporal evolution of species.

Mutations appear in genes at a practically constant rate. For example, in vertebrates the overall molecular evolution takes place at a rate of one amino acid in 2 years. However, the rate of amino acid substitution is different depending on the protein and even on the section of the protein studied. Owing to the degenerate nature of the genetic code, not every point mutation must cause a change in an amino acid in the protein affected. Mutations that are cryptic represent a serious obstacle to an accurate interpretation of molecular evolutionary chronology.

An example of such chronology may be found in the haemoglobin of carp and of man. Mutations in genes coding for haemoglobin accumulated in the genes of the two species at nearly the same rate after the separation of their evolutionary lines about 500 million years ago. Genes for haemoglobin underwent mutations in other species at practically the same rate.

A molecular approach to the study of evolution of species makes it possible to grasp some relationships that would be missed in a classification based solely on a comparison of the phenotype. Thus it was found that differences in the structures of proteins of some frogs of a practically identical phenotype may amount to as much as 20%. Similar comparisons between man and chimpanzee revealed differences of less than 1%. On the basis of this great similarity, it would appear that the evolutionary lines of man and chimpanzee diverged about 5 million years ago whereas palaentologists assumed a period of about 14 million years. The discrepancy between these data can be accounted for either by assuming that the molecular evolution of primates is slower than that of lower vertebrates or by the fact that primates have relatively long generation times.

It may be seen that molecular evolution and the evolution of anatomical features proceeded independently, possibly even through different mechanisms. At any rate, comparisons of a greater number of analogous proteins and the corresponding DNAs in a variety of species will certainly contribute in the future to a much more accurate determination of the time when the individual species started their independent evolution.

Further Reading

More details on the molecular genetics of microorganisms, together with their biochemical and biological aspects, can be found in a great variety of books. More profound information can be found in the following recent English-written publications:

Alberts B., Bray D., Lewis J., Raff M., Roberts K., Watson J. D.: Molecular Biology of the Cell. Garland Publishing, Inc., New York – London 1983.

Annual Reviews of Genetics. Annual Reviews Inc., Palo Alto.

Avers C. J.: Genetics. Revised Edition. D. van Nostrand Company, New York – Cincinnati – Toronto – London – Melbourne 1980.

Ayala F. J., Kiger J. A.: Modern Genetics. The Benjamin/Cummings Publishing Company, Menlo Park – Reading – London – Amsterdam – Don Mills – Sydney 1980.

Bainbridge B. W.: Genetics of Microbes. Blackie, Glasgow – London 1980.

Bielka H. (ed.): The Eukaryotic Ribosome. Akademie-Verlag, Berlin 1982.

Birge E. A.: Bacterial and Bacteriophage Genetics. An Introduction. Springer-Verlag, New York – Heidelberg – Berlin 1981.

Bukhari A. I., Shapiro J. A., Adhya S. L. (ed.): DNA Insertion Elements, Plasmids, and Episomes. Cold Spring Harbor Laboratory, Cold Spring Harbor, New York 1977.

Celis J. E., Smith J. D.: Nonsense Mutations and tRNA Supressors. Academic Press, London – New York – San Francisco 1979.

Davidson J. N.: The Biochemistry of Nucleic Acids, 7th Edition. Academic Press, New York 1972.

Davis R. W., Botstein D., Roth J. R.: Advanced Bacterial Genetics. A Manual of Genetic Engineering. Cold Spring Harbor Laboratory, Cold Spring Harbor, New York 1980.

Day M. J.: Plasmids. Edward Arnold, London 1982.

Denhardt D. T., Dressler D., Ray D. S. (ed.): The Single-Stranded DNA Phages. Cold Spring Harbor Laboratory, Cold Spring Harbor, New York 1978.

Falconer D. S.: Introduction to Quantitative Genetics. 2nd Edition. Longman, New York 1981.

Fincham J. R. S.: Microbial and Molecular Genetics. 2nd Edition. Hodder and Stoughton, London – Sydney – Auckland – Toronto 1976.

Fincham J. R. S.: Genetics. Wright-PSG, Bristol 1983.

Fox S. W., Dose K.: Molecular Evolution and the Origin of Life. Marcel Dekker Inc., New York – Basel 1977.

Friedberg E. C., Hanawalt P. C. (ed.): DNA Repair. A Laboratory Manual of Research Procedures. Marcel Dekker, Inc., New York – Basel 1981.

Glover D. M.: Genetic Engineering – Cloning DNA. Chapman and Hall, London – New York 1980.

Graffi A., Magdon E., Matthes T., Tannenberger S. (ed.): DNA Repair and Cancer Research. Akademie-Verlag, Berlin 1979.

Hanawalt P. C., Haynes R. H. (ed.): The Chemical Basis of Life. Freeman, San Francisco 1973.

Hayes W.: The Genetics of Bacteria and Their Viruses. Studies in Basic Genetics and Molecular Biology. 2nd Edition. Wiley, New York 1969.

Herskowitz I. H.: Principles of Genetics. MacMillan, New York 1973.

Hüttermann J., Köhnlein W., Téoule R., Bertinchamps A.: Effects of Ionizing Radiation on DNA. Physical, Chemical and Biological Aspects. Springer-Verlag, Berlin – Heidelberg – New York 1978.

Kornberg A.: DNA Synthesis. Freeman, San Francisco 1974.
Kornberg A.: DNA Replication. Freeman, San Francisco 1980.
Leaver C. J. (ed.): Genome Organization and Expression in Plants. Plenum Press, New York – London 1980.
Lehninger A. L.: Biochemistry. 2nd Edition. Worth Publishers, Inc., New York 1975.
Lehninger A. L.: Principles of Biochemistry. Worth Publishers, Inc., New York 1982.
Levine L.: Biology of the Gene. The C. V. Mosby Company, St. Louis – Toronto – London 1980.
Levitzki A.: Quantitative Aspects of Allosteric Mechanisms. Springer-Verlag, Berlin – Heidelberg – New York 1978.
Mange A. P., Mange E. J.: Genetics: Human Aspects. Saunders College, Holt, Rinehart and Windston, Philadelphia 1980.
McGilvery R. W.: Biochemistry. A Functional Approach. 2nd Edition. W. B. Saunders Co., Philadelphia – London – Toronto 1979.
Meynell G. C.: Bacterial Plasmids. MacMillan Press, London, MIT Press, Cambridge 1973.
Miller J. H., Reznikoff W. S. (ed.): The Operon. Cold Spring Harbor Laboratory, Cold Spring Harbor, New York 1978.
Morgan J., Whelan W. J. (ed.): Recombinant DNA and Genetic Experimentation. Pergamon Press, Oxford – New York – Toronto – Sydney – Paris – Frankfurt 1979.
Old R. W., Primrose S. B.: Principles of Gene Manipulation. An Introduction to Genetic Engineering. 2nd Edition. Blackwell Scientific Publication, Oxford 1981.
Roy A. K., Clark J. H. (ed.): Gene Regulation by Steroid Hormones. Springer-Verlag, New York – Heidelberg – Berlin 1980.
Schimmel P. R., Söll D., Abelson J. N. (ed.): Transfer RNA: Structure, Properties, and Recognition. Cold Spring Harbor Laboratory, Cold Spring Harbor, New York 1979.
Setlow J. K., Hollaender A. (ed.): Genetic Engineering. Principles and Methods. Plenum Press, New York – London 1981.
Stahl F. W.: Genetic Recombination. Freeman, San Francisco 1979.
Stent G. S.: Molecular Genetics: An Introductory Narrative. Freeman, San Francisco 1971.
Stent G. S., Calendar R.: Molecular Genetics. 2nd Edition. Freeman, San Francisco 1978.
Steward A. D., Hunt D. M.: The Genetic Basis of Development. Blackie, Glasgow – London 1982.
Stryer L.: Biochemistry. 2nd Edition. Freeman, San Francisco 1981.
Székely M.: From DNA to Protein – the Transfer of Genetic Information. The Macmillan Press Ltd., London – Basingstoke 1980.
Tamarin R. H.: Principles of Genetics. Willard Grant Press, Boston 1982.
Tooze J. (ed.): DNA Tumor Viruses. Cold Spring Harbor Laboratory, Cold Spring Harbor, New York 1980.
Vázquez D.: Inhibitors of Protein Biosynthesis. Springer-Verlag, Berlin – Heidelberg – New York 1979.
Watson J. D.: Molecular Biology of the Gene. 3rd Edition. Benjamin-Cummings, Menlo Park 1976.
Watson J. D., Tooze J.: The DNA Story (A Documentary History of Gene Cloning). Freeman, San Francisco 1981.
Whelan W. J., Black S. (ed.): From Genetic Experimentation to Biotechnology. The Critical Transition. Wiley, Chichester 1982.
Woods R. A.: Biochemical Genetics. 2nd Edition. Chapman and Hall, London – New York 1980.

Abbreviations and Symbols

The abbreviations and symbols of nucleic acids, polynucleotides and their components follow the recommendations of the IUPAC-IUB Committee for Biochemical Nomenclature.

Nucleic Acids, Polynucleotides and their Components

Nucleic acids

DNA	deoxyribonucleic acid or deoxyribonucleate
mtDNA	mitochondrial DNA
rDNA	DNA coding for ribosomal RNA
RNA	ribonucleic acid or ribonucleate
cRNA	complementary RNA
hnRNA	heterogeneous nuclear RNA
mRNA	messenger RNA
nRNA	nuclear RNA
rRNA	ribosomal RNA
snRNA	small nuclear RNA
tRNA	transfer RNA
$tRNA^{Ala}$	free transfer RNA for alanine
alanyl-$tRNA^{Ala}$, Ala-$tRNA^{Ala}$	acylated transfer RNA for alanine
$tRNA^{Ala}_{1}$, $tRNA^{Ala}_{2}$	isoacceptor tRNAs for alanine
$tRNA^{Met}$	nonacylated, nonformylatable tRNA for methionine
Met-$tRNA^{Met}$	acylated, nonformylatable tRNA for methionine
$tRNA^{fMet}$	nonacylated, formylatable tRNA for methionine
Met-$tRNA^{fMet}$	acylated, formylatable tRNA for methionine
fMet-$tRNA^{fMet}$	formylacylated, formylatable tRNA for methionine

Nucleotides

AMP	adenosine 5′-monophosphate
cAMP	cyclic adenosine 3′,5′-monophosphate
ADP	adenosine 5′-diphosphate

ATP	adenosine 5′-triphosphate
GMP	guanosine 5′-monophosphate
GDP	guanosine 5′-diphosphate
GTP	guanosine 5′-triphosphate
IMP	inosine 5′-monophosphate
RMP	any unspecified purine ribonucleoside 5′-monophosphate
CMP	cytidine 5′-monophosphate
CDP	cytidine 5′-diphosphate
CTP	cytidine 5′-triphosphate
OMP	orotidine 5′-monophosphate
TMP	ribothymidine 5′-monophosphate
TDP	ribothymidine 5′-diphosphate
TTP	ribothymidine 5′-triphosphate
UMP	uridine 5′-monophosphate
UDP	uridine 5′-diphosphate
UTP	uridine 5′-triphosphate
ΨMP	pseudouridine 5′-monophosphate
ΨDP	pseudouridine 5′-diphosphate
ΨTP	pseudouridine 5′-triphosphate
UDPG, UDPGlc	uridine diphosphate glucose
YMP	any unspecified pyrimidine nucleoside 5′-monophosphate
NMP	any unspecified nucleoside 5′-monophosphate

Nucleotide coenzymes

FMN	riboflavin 5′-monophosphate (flavin mononucleotide)
FAD, $FADH_2$	flavin adenine dinucleotide (oxidized, reduced)
NMN	nicotinamide mononucleotide
NAD^+, NADH	nicotinamide adenine dinucleotide (oxidized, reduced)
$NADP^+$, NADPH	nicotinamide adenine dinucleotide phosphate (oxidized, reduced)

Three-letter symbols

Bases

Ade	adenine
Gua	guanine
Xan	xanthine
Hyp	hypoxanthine
Pur	any unspecified purine
Cyt	cytosine
Ura	uracil

Thy	thymine
Oro	orotate
Pyr	any unspecified pyrimidine

Ribonucleosides

Ado	adenosine
Guo	guanosine
Ino	inosine
Sno	thioinosine
Xao	xanthosine
Puo	any unspecified purine ribonucleoside
Cyd	cytidine
Urd	uridine
Thd	ribosylthymine (not thymidine)
Srd	thiouridine
Ψrd	pseudouridine
Ord	orotidine
Pyd	any unspecified pyrimidine ribonucleoside
Nir	ribosyl nicotinamide

Deoxyribonucleosides

dAdo	2′-deoxyribosyladenine (deoxyadenosine)
dGuo	2′-deoxyribosylguanine (deoxyguanosine)
dPuo	any unspecified 2′-deoxyribosylpurine
dUrd	2′-deoxyribosyluracil (deoxyuridine)
dCyd	2′-deoxyribosylcytosine (deoxycytidine)
dThd	2′-deoxyribosylthymine (thymidine)
dPyd	any unspecified 2′-deoxyribosylpyrimidine

Nucleotides

Ado-5′-P (P-5′-Ado)	adenosine 5′-monophosphate
Guo-5′-P (P-5′-Guo)	guanosine 5′-monophosphate
Urd-5′-P (P-5′-Urd)	uridine 5′-monophosphate
Cyd-5′-P (P-5′-Cyd)	cytidine 5′-monophosphate
Ado-3′:5′-P(P-3′:5′-Ado)	cyclic adenosine monophosphate
Ado-3′,5′-P_2(P_2-3′,5′-Ado)	adenosine 3′,5′-bisphosphate
Ado-5′-PP	adenosine 5′-diphosphate
dAdo-5′-P	2′-deoxyadenosine 5′-monophosphate
dThd-5′-P	thymidine 5′-monophosphate
Ado-3′-P 5′-Urd	adenosyluridine
Urd-5′-PP-Glc	uridine diphosphate glucose

One-letter symbols

Ribonucleosides

A	adenosine
G	guanosine
I	inosine
M	6-thioinosine (ribosylmercaptopurine)
X	xanthosine
R	any unspecified purine ribonucleoside
C	cytidine
U	uridine
Ψ	pseudouridine
T	ribosylthymine (not thymidine)
Y	any unspecified pyrimidine ribonucleoside
N	any unspecified ribonucleoside
D	5,6-dihydrouridine
S	thiouridine
B	5-bromouridine
O	orotidine

Deoxyribonucleosides

dA	2′-deoxyribosyladenine (deoxyadenosine)
dG	2′-deoxyribosylguanine (deoxyguanosine)
dR	any unspecified 2′-deoxyribosylpurine
dU	2′-deoxyribosyluracil (deoxyuridine)
dC	2′-deoxyribosylcytosine (deoxycytidine)
dT	2′-deoxyribosylthymine (thymidine)
dY	any unspecified 2′-deoxyribosylpyrimidine

Nucleotides

Ap	adenosine 3′-monophosphate
pA	adenosine 5′-monophosphate
pAp	adenosine 3′,5′-bisphosphate
pppA	adenosine 5′-triphosphate
ppGpp	guanosine tetraphosphate [guanosine bis(diphosphate)]

Oligonucleotides

G3′p5′A3′p5′A G3′-5′A3′-5′A (3′→5′)G-A-A G-A-A	guanylyladenylyladenosine with 3′→5′ internucleotide bonds

(2′→5′)G-A-A	guanylyladenylyladenosine with 2′→5′ internucleotide bonds
G-A-Ap	guanylyladenylyladenosine with a 3′-terminal phosphate
G-A-A>p	guanylyladenylyladenosine with 2′,3′-cyclic phosphate
d(G-A-A)	deoxyguanylyldeoxyadenylyldeoxyadenosine with 3′→5′ internucleotide bonds
d(A←A←G)	the same oligodeoxynucleotide (the arrow indicates a 5′←3′ internucleotide bond)
GAA	a G-A-A codon; for codon abbreviations the symbols for phosphoric acid (- or p) are omitted

Polynucleotides

poly(A)	polyadenylate
poly(A-C)	poly(adenylate-cytidylate) with regularly alternating nucleotides
poly(A,C)	poly(adenylate,cytidylate) with random sequence of the two nucleotides
poly(A_2,C)	as above, containing a 2 : 1 ratio of A to C
poly(dA-dT),poly[d(A-T)]	polymer with alternating dA and dT
poly(A) . poly(U)	noncovalently bound chains of poly(A) and poly(U) where A always pairs with U, irrespective of chain length*
poly[d(A-T)] . poly[d(A-T)] poly[d(A-T). d(A-T)]	noncovalently bound two polydeoxyribonucleotides

Amino Acids

Ala	alanine
Arg	arginine
Asn	asparagine
Asp	aspartic acid
Asx	aspartic acid or asparagine
Cys	cysteine
Gln	glutamine
Glu	glutamic acid
Glx	glutamic acid or glutamine
Gly	glycine
His	histidine

---

* The central dot indicates pairs of bases participating in a noncovalent bond; A·T or G·C represent hydrogen bonds between the bases whereas A-T or G-C represent covalent bonds.

Ile	isoleucine
Leu	leucine
Lys	lysine
Met	methionine
Phe	phenylalanine
Pro	proline
Ser	serine
Thr	threonine
Trp	tryptophan
Tyr	tyrosine
Val	valine

Index

If a term appears frequently in the book, only the numbers of those pages are given, which involve significant amounts of text concerning the term.